Revolutions in Mathematics

Problem Solving in Mathematics and Beyond

Print ISSN: 2591-7234
Online ISSN: 2591-7242

Series Editor: Dr. Alfred S. Posamentier
Distinguished Lecturer
New York City College of Technology - City University of New York

There are countless applications that would be considered problem solving in mathematics and beyond. One could even argue that most of mathematics in one way or another involves solving problems. However, this series is intended to be of interest to the general audience with the sole purpose of demonstrating the power and beauty of mathematics through clever problem-solving experiences.

Each of the books will be aimed at the general audience, which implies that the writing level will be such that it will not be engulfed in technical language — rather the language will be simple everyday language so that the focus can remain on the content and not be distracted by unnecessarily sophiscated language. Again, the primary purpose of this series is to approach the topic of mathematics problem-solving in a most appealing and attractive way in order to win more of the general public to appreciate this most important subject rather than to fear it. At the same time we expect that professionals in the scientific community will also find these books attractive, as they will provide many entertaining surprises for the unsuspecting reader.

Published

Vol. 44 *Revolutions in Mathematics*
by Alfred S Posamentier and Arthur D Kramer

Vol. 43 *Mathematical Puzzles and Curiosities*
by Ivo Fagundes David de Oliveira, Tanya Khovanova and
Yogev Shpilman

Vol. 42 *Learning Geometry Through Problem Solving*
by Boris Pritsker

Vol. 41 *Maxima and Minima: How Extremes Can Explain Mathematics
Concepts and Facilitate Problem Solving*
by Hans Humenberger and Alfred S Posamentier

For the complete list of volumes in this series, please visit www.worldscientific.com/series/psmb

Problem Solving in Mathematics and Beyond

Volume **44**

Revolutions in Mathematics

Arthur D. Kramer
The City University of New York, USA

Alfred S. Posamentier
The City University of New York, USA

World Scientific

NEW JERSEY · LONDON · SINGAPORE · BEIJING · SHANGHAI · HONG KONG · TAIPEI · CHENNAI · TOKYO

Published by

World Scientific Publishing Co. Pte. Ltd.

5 Toh Tuck Link, Singapore 596224

USA office: 27 Warren Street, Suite 401-402, Hackensack, NJ 07601

UK office: 57 Shelton Street, Covent Garden, London WC2H 9HE

Library of Congress Cataloging-in-Publication Data
Names: Kramer, Arthur D., 1940–   author | Posamentier, Alfred S. author
Title: Revolutions in mathematics / Arthur D. Kramer, The City University of New York, USA,
    Alfred S. Posamentier, The City University of New York, USA.
Description: New Jersey : World Scientific, [2026] | Series: Problem solving in mathematics
    and beyond, 2591-7234 ; volume 44 | Includes index.
Identifiers: LCCN 2025054727 | ISBN 9789819825523 hardcover |
    ISBN 9789819825981 paperback | ISBN 9789819825530 ebook |
    ISBN 9789819825547 ebook other
Subjects: LCSH: Mathematics--History
Classification: LCC QA21 .K699 2026
LC record available at https://lccn.loc.gov/2025054727

British Library Cataloguing-in-Publication Data
A catalogue record for this book is available from the British Library.

For any available supplementary material, please visit
https://www.worldscientific.com/worldscibooks/10.1142/14653#t=suppl

Desk Editors: Kannan Krishnan/Gabriel Rawlinson

Typeset by Stallion Press
Email: enquiries@stallionpress.com

*To my dear Donna and her wonderful
family for enriching my life and
bringing me much love and joy.*
Arthur D. Kramer

About the Authors

Arthur D. Kramer honed his youthful years in a wonderful Brooklyn, New York community, moving on to Stuyvesant High School in Manhattan and then earning a BME degree from Cooper Union Engineering College. This was followed by an M.A. in Math Education from Columbia University and a Ph.D. in Mathematics from New York University. After working as an engineer for several years he found his calling in teaching mathematics, first in high school and then at New York University. This led to a life-long career as a professor at New York City College of Technology of the City University of New York. He taught a variety of the mathematics courses, with a specialty in the history and culture of mathematics. He was also invited to teach courses in the Physics, Electrical Engineering and Computer Engineering departments. During this time, he spent several years as an adjunct Professor of Mathematics at the New School for Social Research.

While early on he was interested in mathematical research and published several articles in mathematical journals, he found his real calling in educational pursuits resulting in grants and special programs at New York City College of Technology. In particular, he developed a curriculum for a program called "Access for Women," which held weekend sessions to expose high school girls to various STEM fields. They studied and performed basic experiments in different engineering disciplines with the hope that it would motivate them to enter one of these fields.

Professor Kramer has published five textbooks in technical mathematics, his latest being *Mathematics for Electricity and Electronics* and a recent book with Dr. Alfred S. Posamentier for a general audience entitled *Mathematics: Its Historical Aspects, Wonders and Beyond*. He continues to teach as an adjunct in the Electrical Engineering department at NYC College of Technology.

When not devoting time to education and writing, he is an accomplished offshore sailor and has spent summers sailing his Ketch *Sweet Harmony* along the US East Coast and Nova Scotia. He has made four passages to Bermuda and spent two years sailing across the Atlantic, throughout the Mediterranean and the Caribbean. He enjoys ballroom dancing and has recently discovered a pleasurable pastime playing tennis.

Alfred S. Posamentier is currently a Distinguished Lecturer at New York City College of Technology of the City University of New York. Prior to that he was Executive Director for Internationalization and Funded Programs at Long Island University, New York. This was preceded by 5 years as Dean of the School of Education and Professor of Mathematics Education at Mercy University, New York. Before that he was for 40 years at the City College of the City University of New York, at which he is now Professor Emeritus of Mathematics Education and Dean Emeritus of the School of Education. He is the author and co-author of more than 90 mathematics books for teachers, secondary and elementary school students, as well as the general readership. Dr. Posamentier is also a frequent commentator in newspapers and journals on topics related to education and mathematics.

After completing his B.A. degree in mathematics at Hunter College of the City University of New York, he took a position as a teacher of mathematics at Theodore Roosevelt High School (Bronx, New York), where he focused his attention on improving the students' problem-solving skills and at the same time enriching their instruction far beyond what the traditional textbooks offered. During his six-year tenure there, he also developed the school's first mathematics teams (both at the junior and senior level). He is still involved in working with mathematics teachers and supervisors, nationally and internationally, to help them maximize their effectiveness.

Immediately upon joining the faculty of the City College of New York in 1970 (after having received his master's degree there in 1966), he began to develop in-service courses for secondary school mathematics teachers, including such special areas as recreational mathematics and problem solving in mathematics. As Dean of the City College School of Education for 10 years, his scope of interest covered the full gamut of educational issues. During his tenure as Dean, he took the School of Education from the bottom of the New York State rankings to the top with a perfect NCATE accreditation assessment in 2009. He also raised more than $12,000,000 from the private sector for innovative educational programs. Posamentier repeated this successful transition at Mercy College, where he enabled it to become the only college to have received both NCATE and TEAC accreditation simultaneously.

In 1973, Dr. Posamentier received his Ph.D. from Fordham University (New York) in mathematics education, and he has since extended his reputation in mathematics education to Europe. He has been visiting professor at several European universities in Austria, England, Germany, the Czech Republic, Turkey and Poland. In 1990, he was Fulbright Professor at the University of Vienna.

In 1989 he was awarded an Honorary Fellow position at the South Bank University (London, England). In recognition of his outstanding teaching, the City College Alumni Association named him Educator of the Year in 1994, and in 2009. New York City had the day, May 1, 1994, named in his honor by the President of the New York City Council. In 1994, he was also awarded the *Grosse Ehrenzeichen für Verdienste um die Republik Österreich* (Grand Medal of Honor from the Republic of Austria), and in 1999, upon approval of Parliament, the President of the Republic of Austria awarded him the title of University Professor of Austria. In 2003, he was awarded the title of *Ehrenbürgerschaft* (Honorary Fellow) of the Vienna University of Technology, and in 2004 he was awarded the *Österreichisches Ehrenkreuz für Wissenschaft & Kunst 1. Klasse* (Austrian Cross of Honor for Arts and Science, First Class) from the President of the Republic of Austria. In 2005, he was inducted into the Hunter College Alumni Hall of Fame, and in 2006 he was awarded the prestigious Townsend Harris Medal by the City College Alumni Association. He was inducted into the New York State Mathematics Educators' Hall of Fame in 2009, and in 2010 he was awarded the coveted Christian-Peter-Beuth Prize from the Technische Fachhochschule – Berlin. In 2017, Posamentier was awarded *summa cum laude nemine discrepante* by the Fundacion Sebastian, A.C., Mexico City, Mexico.

He has taken on numerous important leadership positions in mathematics education locally. He was a member of the New York State Education Commissioner's Blue Ribbon Panel on the Math-A Regents Exams, and the Commissioner's Mathematics Standards Committee, which redefined the Mathematics Standards for New York State, and he also served on the New York City Schools' Chancellor's Math Advisory Panel.

Dr. Posamentier is still a leading commentator on educational issues and continues his long-time passion of seeking ways to make mathematics interesting to teachers, students and the general public – as can be seen from some of his more recent books.

For more information and a list of his publications see: https://en.wikipedia.org/wiki/Alfred_S._Posamentier.

Contents

About the Authors   vii

Introduction   xiii

Chapter 1   The Dawn of Mathematics   1
Early Mathematical Achievements

Chapter 2   The Ancient Greeks and Romans   39
The Beginning of Theoretical Mathematics

Chapter 3   The Emergence of Our Number System   55
The Development of the Place-Value Decimal System

Chapter 4   The Origin and Growth of Algebra   77
Algebra Blossoms in the Medieval Era
and the Renaissance

Chapter 5   The Foundations of Probability Theory   105
The Laws of Chance Become Formalized

Chapter 6   The Discovery of Calculus   121
Unraveling the Paradox of the Infinite
and the Infinitesimal

Chapter 7   The Development of Group Theory   149
A Universal Foundation Emerges for Mathematical Systems

Chapter 8   The Breakthrough of Non-Euclidean Geometry   169
A Geometrical Awakening Abdicates Euclidean Geometry

Chapter 9   The Theory of Sets   191
A Basis for Modern Mathematics Becomes "The New Math" of the Twentieth Century

Chapter 10   The Digital Age   203
The Relentless Advance of Computational Speed and Artificial Intelligence

Index   225

Introduction

Revolutions in Mathematics reveals the captivating stories behind major mathematical discoveries that shape our world today. The book is designed for a general audience that desires to understand how our mathematical world has evolved and the human genius that made it possible. What is necessary is only a willingness to engage with mathematical ideas, which are presented in a clear, easily understood manner, stripping away the difficulties that frustrate many readers. The material is made appealing by enhancing it with the intriguing historical events surrounding the mathematical breakthroughs and the underlying concepts. The rich historical background that surrounds these mathematical developments fosters an appreciation for the intense human effort and sacrifice involved and the inevitable conflicts that accompanied them. The book strives to fulfill what is much needed in today's world: a more positive feeling toward mathematics and its ubiquitous applications. By doing so, it is hoped that it will inspire one's curiosity to seek out and learn more about mathematics and science.

Revolutions in Mathematics focuses on ten of the major advancements in mathematics over the last 4,000 years. These ideas form the basis for much of today's science and technology. By presenting these remarkable contributions within their historical context, it should pique the reader's interest, not just in history, but also in the mathematical ideas that emerged. The mathematical discoveries emanated from many brilliant people whose lives and interactions are a very engaging story that embellishes the development of these discoveries. There are doubts, bitter jealousies, fierce competitions, and human despair that all play a role in the discovery and introduction of major new ideas. When many of these ideas were first introduced, they encountered resistance to being accepted. They not only threatened past beliefs but contradicted contemporary theory. Eventually, opposition to these truths diminished as these new concepts proved viable, and they became part of the mainstream.

The first chapter begins with the remarkable achievements of the ancient civilizations that flourished in Mesopotamia and other cradles of civilization many thousands of years before the Common Era and left their footprint on today's mathematical world. This is followed by the early Greek civilization and its considerable contribution to plane geometry. Centuries later, in our modern world the geometry of the Greeks remains an integral part of the school education curriculum. Euclidean geometry not only prevails today but dominated the mathematical world as the only geometry for over a thousand years until the Renaissance spawned the discovery of calculus. The third chapter reveals the origin of our decimal system, which seems so simple and easy to work with, but it took thousands of years to develop, and went through many iterations before it reached its very useful present form.

Once the number system evolved, it gave way during the Renaissance to the development of algebra which the fourth chapter discusses. Algebra facilitated the solution of many problems, and gave rise to more abstract applications. The fifth chapter focuses on one of these important applications, that of probability and statistics, which came about while trying to predict gambling odds and has grown to become very important and essential today in science, medicine, and politics.

The discovery of calculus in the next chapter is a truly fascinating story filled with international political and cultural controversy involving the rivalry of two countries, England and Germany. The reader will gain a clear insight into what calculus is all about by easy-to-follow examples. The seventh chapter discusses a topic generally limited to higher mathematics, but its importance cannot be overlooked, even by the general readership. The Theory of Groups emerged during a tumultuous period in European history, culminating in the loss of a two mathematical geniuses at very young ages. The history of its discovery is equally as fascinating as the subject itself, whose fundamental ideas can be readily grasped. The eighth chapter is truly captivating as it deals with a topic that had its origins in antiquity but took many years before it was fully understood by mathematicians, and well before it could be revealed and accepted by the public. This was because the introduction of non-Euclidean geometry threatened the uniqueness of Euclidean geometry, which was as sacrosanct as the Bible for over a thousand years.

The Theory of Sets presented in the ninth chapter provided a basic structure for all of mathematics when it evolved in the late nineteenth and early twentieth centuries. It was also instrumental in giving rise to the "New Math" curriculum of the 1960s, which was imposed for a time upon the school curricula in the United States. However, New Math was found to emphasize theory at the expense of mastering skills, and eventually it was eliminated from basic education, but set theory remains fundamental to the study of all advanced mathematics today. We conclude with the tenth chapter, which explores our present digital age and the exponential increase in mathematical applications that portend major changes in our world today, some of which may drastically change our quality of life.

We invite you to embark upon this journey into the history of major mathematical discoveries with a desire to better comprehend fundamental ideas that shape the state of our world today.

Chapter 1

The Dawn of Mathematics

We are currently living in an extraordinarily prolific scientific age. At times, we experience such an exponential increase in technical accomplishments that we tend to lose sight of the amazing achievements of past civilizations and possibly relegate them to a time of limited scientific advancement. Nothing could be further from the truth, as human capability and ingenuity are not any greater today than they have ever been. One only needs to look at the astounding accomplishments of the ancient Egyptians, and the pyramids they built more than four thousand years ago and their incredible tombs constructed many stories below the earth. These structures, both above and below ground, continue to confound our understanding of how and when they were built. The Egyptians' knowledge of geometry and mathematics clearly played a major role in their achievements, which were accomplished with far less sophisticated mathematics and technology than those that exist today. Other early civilizations of that time were also advanced in their urban architecture, water systems, agriculture, science, and astronomy. Between 4000 BCE and 3000 BCE, these civilizations emerged in present-day Egypt, Iran, Iraq, India, China, Peru, and Mexico, and made significant contributions that still endure today.

The Sumerians

A significant birthplace of civilization is Mesopotamia, which means "between two rivers" in Greek, and is located in what is today Iraq, Kuwait, and Syria. Although the evidence we have unearthed is not definitive, it appears that the earliest major advances in mathematics date back to the Sumerian civilization, which flourished from 4100 to 2000 BCE between the Tigris and Euphrates rivers on the edge of the Persian Gulf. (See Figure 1.1.) The Sumerians, along with the ancient Egyptians and the Elamites in present-day Iran, lived in one of three of the cradles of civilization that flourished in the Western world. Inhabiting the rich valleys of the Tigris and Euphrates rivers, the Sumerian farmers produced a plentiful supply of grains and other crops. This resulted in urban settlements creating a primitive form of writing between 3500 BCE and 3000 BCE.

Figure 1.1[1] Sumerian civilization 4100–2000 BCE.

[1] Property of Nadhir Al-Ansari.

It was the Sumerians who inspired some of the mathematics that we still embrace today. The oldest mathematical evidence we have is a clay tablet dating to around 2700 BCE that contains a multiplication table of areas, involving the product of two lengths of a figure, most likely a rectangle. The tablet has a small square surface of about 3 inches on each side and is less than an inch thick (Figure 1.2).

Figure 1.2   Oldest Sumerian tablet circa 2700 BCE.

The writing is in cuneiform, which uses wedge-like characters. A scribe used a stylus to imprint the characters on a wet clay tablet, which was then dried in the Sun or in kilns, making them very durable and able to last for years. Many thousands of tablets have been found in the Middle East. The first major discovery was made in 1850 by the English explorer Sir Austin Henry Layard (1817–1894) (Figure 1.3) in Nineveh, which is in present-day Mosul, Iraq, and was the largest city in the world for almost 50 years. In 1974–75, a team of Italian archeologists discovered more than 17,000 clay tablets and fragments in Tell Mardikh, a village in Syria.

We know that the Sumerians divided the year into 30-day months, most likely because the cycle of the moon repeats every 29.5 days. As a result, they had 12 months and 360 days in their year. The numbers 12 and 30, inspired by the lunar cycle, may have been the reason why

Figure 1.3   Sir Austin Henry Layard.

they divided the day into 12 periods, where each period corresponded to 2 of our hours, and divided each period into 30 parts, where each part corresponded to 4 of our minutes. What is mathematically interesting is that they chose 60 for the base of their number system, creating a sexagesimal system rather than a base 10 or decimal system. It is not clear why the Sumerians adopted such a base and did not choose a base of ten, which stems from our ten fingers. One reasonable assumption, which we speculate could be true, is the ease of computation that a base of 60 provides. This is because the number 60 possesses all of the following divisors: 2, 3, 4, 5, 6, 10, 12, 15, 20, and 30, as opposed to 10, which only has the divisors 2 and 5. Furthermore, 60 divides into 360, the days in their year. This property of 60 would have enabled the Sumerians to more easily perform arithmetic processes with fractions and find common denominators, which was an important skill needed for the practical mathematics of that era. For example, suppose you wanted to add the following fractions:

$$\frac{4}{15}+\frac{5}{12}$$

The common denominator is 60. The fractions can be simply converted to denominators of 60 and added as follows:

$$\frac{16}{60}+\frac{25}{60}=\frac{41}{60}$$

A Sumerian scholar would most likely have committed the factors of 60 to memory and performed the addition mentally, in the same way we are taught to memorize our current multiplication tables. How would you mentally add the following fractions using a denominator of 60:

$$\frac{3}{10}+\frac{4}{15}$$

You would proceed as follows:

$$\frac{3}{10}+\frac{4}{15}=\frac{18}{60}+\frac{16}{60}=\frac{34}{60}$$

The Sumerian fractions having a denominator of 60 are analogous to our decimal fractions having a denominator of 10. Another number that is also a perfectly acceptable number base is 12. It is easier to calculate in base 12 than base 10 because 12 has four divisors: 2, 3, 4, and 6, as opposed to the two divisors, 2 and 5, of base 10. However, there is no historical evidence of any civilization that used 12 as a number base.

The Sumerians had a lasting effect on the mathematics of future civilizations because they were an advanced civilization that built cities, schools, and taverns, which were public eating places where people could gather for food or drink. They had a governmental bureaucracy, with an actual postal service, were accomplished in architecture, and designed and built systems of irrigation.

The Babylonians

Around the year 2000 BCE, the Semitic Babylonians invaded Mesopotamia and defeated the Sumerians. They established their capital at Babylon around 1900 BCE. (See Figure 1.4.) The Sumerian civilization became completely unknown, and we only learned about it when artifacts were surprisingly unearthed during the mid-19th century with excavations in Mesopotamia. Since the 1850s, some 400 clay tablets have been unearthed, which have revealed a substantial amount of Babylonian mathematics. Most of the tablets that were found date from 1800 BCE to 1600 BCE. However, a substantial

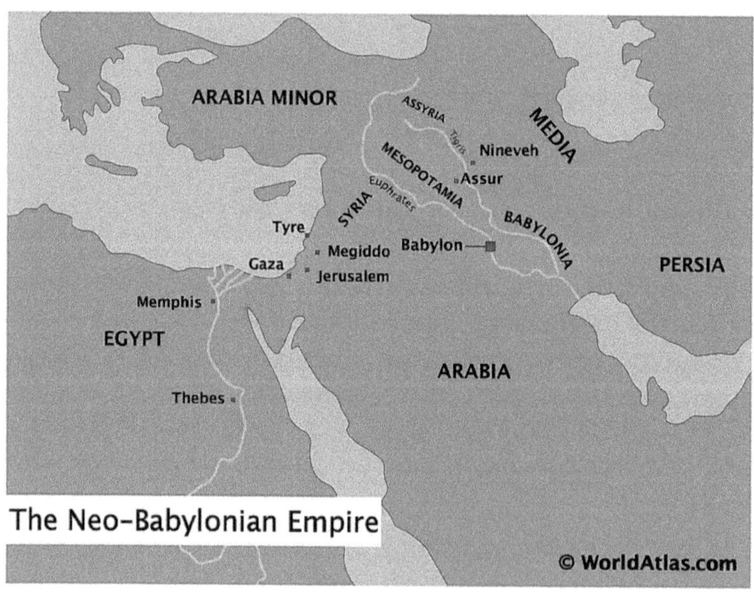

Figure 1.4   Babylonian Empire.

number of them remain to be deciphered. A major discovery has recently been revealed on one of them, Plimpton 322 (Figure 1.5), and is discussed below at the end of this section. The Babylonians drew upon the mathematics of the Sumerians and used the base of 60, but went on to create a more advanced mathematical system. Their abstract form of writing was very similar to the cuneiform, or wedge-shaped, symbols used by the Sumerians.

Have you ever wondered why the day is divided into 24 hours, the hour into 60 minutes, and the minute into 60 seconds? It was the Babylonians who set aside 12 hours of daytime, 12 hours of nighttime, and divided the hours and minutes by 60 because of their sexagesimal system. Miraculously, these concepts survived for over six thousand years, are used throughout the world, and will no doubt remain with us. If you were to set the number of hours in the day and the number of minutes in the hour, based on our decimal system, what would be the logical choices? Most likely, 20 hours in the day would be divided into 10 hours of daytime and 10 hours of nighttime, with each hour being longer than our current hour and containing either 100 or 50 longer minutes.

Figure 1.5   Plimpton 322, an ancient Mesopotamian mathematical tablet.

Here is a further look at how the sexagesimal system works. Consider the time: 3 hours, 20 minutes, and 10 seconds. This corresponds to the number: $3 + \frac{20}{60} + \frac{10}{3600}$. Note that the denominators of the two fractions are multiples of 60. Therefore, in a base 60 system, the Babylonians could simply write this time as 3,20,10. We use the commas here as *sexagesimal points*. In this notation, the $20 = \frac{20}{60}$, and it corresponds to our decimal place of $0.1 = \frac{1}{10}$. The $10 = \frac{10}{3600}$, and it corresponds to our decimal place of $0.01 = \frac{1}{100}$. The sexagesimal point is analogous to the decimal point used in our base 10 system and it represents fractions whose denominators are multiples of 60. Therefore, in the Babylonian number system, the time of day can be simply represented as a sexagesimal number.

This representation of time was accepted by almost every civilization since the Babylonians, which leads one to wonder that perhaps the world was comfortable with base 60 time. Does base 60 time possess a somewhat natural feeling, or are we just so accustomed to denoting time this way that it has become natural? Whatever the reason, Babylonian time measurement was a truly remarkable ancient mathematical contribution that all of us take for granted and rarely question. Sexagesimal time divisions became universal mainly because of the advanced Babylonian civilization. However, we now know it was the Sumerians who first employed the base 60 system and that the Babylonians later adopted it. Their daily division of 24 hours was different than the Sumerian division of 12 periods or hours.

Whereas the Sumerian system was not an advanced positional number system like ours, which has place values of 1, 10, 100 (10^2), 1000 (10^3), etc., the Babylonian number system was, remarkably, a positional number system with place values of 1, 60, 360 (60^2), 21,600 (60^3) etc. This was an extraordinary achievement given that no other known civilization employed a positional number system until thousands of years later, when a system like ours began to develop and take root, evolving into our present number system. The Egyptians, Greeks, and Romans did not use a positional number system, and Roman numerals dominated for over 1500 years. As a result of their number system, the Babylonians were able to do complex calculations to solve real-life problems, such as designing elaborate canals for irrigation and producing amazing architecture. There were just two basic symbols used in their number system, a one and a ten. All the numbers were constructed by repeating these two symbols. Figure 1.6 shows some Babylonian numbers using these two symbols.

Figure 1.6   Babylonian numerals.

It is interesting to note that though their number system was based on 60, the number 10 did have significance in their system. We could then assume that their fingers may have played a part in their arithmetic. Note in Figure 1.6 that the numbers 2 and 61 are written with the same two characters, and they could cause confusion. However, the two characters touch each other in the number 2, indicating one place value, while the space between the symbols for 61 represents two place values. The first-place value on the right is the 1's position, while the second-place value to the left is the 60's position. The next place values to the left are then 60^2, 60^3, etc. For example, consider the number:

$$3924 = 3600 + 300 + 33 = 1(60^2) + 5(60) + 33(1)$$

In base 60, we could write 3924 as **1, 5, 33**, where the commas separate the place values. In Babylonian cuneiform, it is:

How would you write the number 3681 in the Babylonian number system?

It would look like:

$$3681 = 1(3600) + 1(60) + 21(1) = 1, 1, 21 =$$

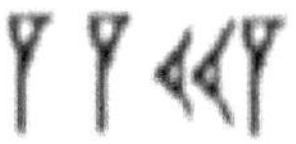

There were some shortcomings with this system because there was no symbol for zero. This is not surprising, as many ancient civilizations did not see the need for a number with no value. As a result, another possible confusion could arise between the numbers 1 and 60, which are both written with the same single character. We can only assume that somehow the Babylonians were able to distinguish the difference, perhaps within the context in which the numbers appeared. Their number skills were certainly sufficient to eliminate this confusion, so they may not have considered it a difficult problem. Later Babylonian civilizations did invent a symbol for zero, but it was only used between characters with different place values to avoid misinterpretation. For example, the number 3601 would require a space in the 60's position between the 1 and the 3600 (60^2) position. If there was no symbol for zero, the symbol for 3600 in the third place could be interpreted as the number 60 in the second place, resulting in the number 61. However, a scribe might leave a double space between the two positions to avoid this confusion. There still was no symbol created for zero that could be used in the unit's position, so a number like 60 could be misinterpreted.

The Babylonians also worked with fractions whose denominators were the sexagesimal values, 60, 3600 (60^2), 216,000 (60^3), etc., which were analogous to our decimal fractions with denominators of 10, 10^2, 10^3, etc. Now that may seem difficult, but since 60 has more divisors than 10, more of their fractions could be represented as finite decimals than fractions in the base 10 system. Here are some examples. The fraction $\frac{1}{5}$ can be written exactly as a decimal fraction $\frac{2}{10} = 0.2$ and exactly as a sexagesimal fraction $\frac{12}{60} = ,12$. Here again, the comma is used as a sexagesimal point analogous to the decimal point. The fractions $\frac{1}{3} = 0.333...$ and $\frac{1}{9} = 0.111...$ are infinite decimals and cannot be written as decimal fractions. However, they can be written as sexagesimal fractions as follows:

$$\frac{1}{3} = \frac{20}{60} = ,20 \quad \text{and} \quad \frac{1}{9} = \frac{400}{3600} = \frac{360}{3600} + \frac{40}{3600} = \frac{6}{60} + \frac{40}{3600} = ,6,40$$

Similarly, $\frac{1}{8}$ as a decimal is 0.125. It can be written as a sexagesimal fraction as follows:

$$\frac{1}{8} = \frac{450}{3600} = \frac{420}{3600} + \frac{30}{3600} = \frac{7}{60} + \frac{30}{3600} = ,7,30$$

The reader is welcome to check the correctness of the arithmetic in the above examples.

Figure 1.7 compares eight basic decimal fractions with sexagesimal fractions. Notice that three of the fractions, $\frac{1}{3}$, $\frac{1}{6}$ and $\frac{1}{9}$ can not be represented as finite decimal fractions, but they can be represented as finite sexagesimal fractions.

System	$\frac{1}{2}$	$\frac{1}{3}$	$\frac{1}{4}$	$\frac{1}{5}$	$\frac{1}{6}$	$\frac{1}{8}$	$\frac{1}{9}$	$\frac{1}{10}$
Decimal	0.5		0.25	0.20		0.125		0.10
Sexagesimal	,30	,20	,15	,12	,10	,7,30	,6,40	,6

Figure 1.7   Decimal fractions vs sexagesimal fractions.

Another example is the fraction $\frac{1}{150}$, which cannot be written as a decimal fraction but can be written as the sexagesimal fraction $\frac{24}{3600} = ,,24$ (double comma). What would the fractions $\frac{1}{12}$ and $\frac{1}{120}$ look like as sexagesimal fractions?[2] The reader is encouraged to try to represent other common fractions in the sexagesimal system.

The Babylonians, like the Sumerians, divided the circle into 360 degrees (360°), which is a multiple of their base 60. We have come to universally accept it as a basic mathematical measurement. It has been suggested that this might have been motivated by the number of

---

[2] $\frac{1}{12} = \frac{5}{60} = ,5$ and $\frac{1}{120} = \frac{3}{3600} = ,,3$ (double comma).

days in the year. However, it is reasonable to assume, given their astronomical expertise, that the ancients knew this was not true. Here we quote from an article by J.J. O'Connor and E.F. Robertson about number bases:

> *I just do not believe that anyone ever chose a number base for any civilization. Can you imagine the Sumerians setting up a committee to decide on their number base. Things just did not happen in that way. The reason has to involve the way that counting arose in the Sumerian civilization, just as 10 became a base in other civilizations who began counting on their fingers, and twenty became a base for those who counted on both their fingers and toes.*
>
> *Here is one way that it could have happened. One can count up to 60 using your two hands. On your left hand there are three parts on each of four fingers (excluding the thumb). The parts are divided from each other by the joints in the fingers. Now one can count up to 60 by pointing at one of the twelve parts of the fingers of the left hand with one of the five fingers of the right hand. This gives a way of finger counting up to 60 rather than to 10. Anyone convinced?*[3]

In this supposition by O'Connor and Robertson, the twelve parts on the four fingers of the left hand are counted five times by using the five fingers of the right hand. Therefore, 5 times 12 equals 60. Perhaps you have your own idea about using 60 as a base.

Dividing the circle into 360° bodes well geometrically as follows. The sum of the angles of any triangle is then proven to be a straight angle = 180°, making each angle of an equilateral triangle 60°. If we were to have been motivated by our decimal system, we would likely have assigned 400° to the circle. Then each quadrant would be 100° instead of 90°. However, the sum of the angles of any triangle would then be a straight angle = 200°, and the angles of an equilateral triangle would each have to be $\frac{200}{3} = 66\frac{2}{3}$ degrees. Not an easy value to assign to such an important triangle. This leads us to appreciate further the contributions of the Babylonians and their sexagesimal system.

---

[3] https://mathshistory.st-andrews.ac.uk/HistTopics/Babylonian_numerals/

Perhaps the most impressive calculating skills of the Babylonians were found in their creation of extensive tables used to simplify computation. Tablets that date from 2000 BCE were found in 1854 and contain squares of numbers up to 59^2 and cubes up to 32^3. This table used a very clever technique to multiply numbers. The following examples require some algebra and can be read lightly, but will still highlight the arithmetic ingenuity of the Babylonians. To multiply two numbers a and b, they used the formula:

$$a \times b = \tfrac{1}{4} \times [(a+b)^2 - (a-b)^2]$$

Using some algebra, we show that this formula works:

$$\frac{1}{4}\left[(a+b)^2 - (a-b)^2\right] = \frac{1}{4}\left[\left(a^2 + 2ab + b\right)^2 - \left(a^2 - 2ab + b\right)^2\right]$$

$$= \frac{1}{4} \times [4ab] = ab$$

For example, to multiply 16×36, they would first obtain the squares of the sum and the squares of the difference of these numbers from the table. Then they would subtract the two values and take one-fourth of the result to obtain the product. The calculation in base 10 is as follows:

$$16 \times 36 = \frac{1}{4}\left[(16+36)^2 - (36-16)^2\right] = \frac{1}{4}\left[52^2 - 20^2\right]$$

$$= \frac{1}{4}[2704 - 400] = \frac{2304}{4} = 576$$

The Babylonians, of course, would do this problem in base 60. You may want to apply the formula to 24×18 and show that the product is 432. Given today's technology, this may seem tedious, but 4000 years ago it was a significant achievement. Division was even more challenging, but the Babylonians performed it also in a very clever way. They did not have an algorithm for long division, but they realized that division is the same as multiplying by the reciprocal of the divisor:

$$a \div b = a \times \frac{1}{b}$$

They constructed an incredibly large table of reciprocals up to several billion and reduced division to multiplication. Their reciprocals were, of course, sexagesimal fractions. Consider a simple division example where we will use $\frac{1}{3} = \frac{20}{60}$ in base 60:

$$15 \div 3 = 15 \times \frac{1}{3} = 15 \times \frac{20}{60} = 15 \times ,20 = ,300 = \frac{300}{60} = 5$$

Another division problem is $160 \div 20$. Since $160 = 2(60) + 40$, we write the number 160 as 2, 40, and the division would look something like:

$$160 \div 20 = 160 \times \frac{1}{20} = [2, 40] \times \frac{3}{60} = [2, 40] \times ,3$$

$$= \frac{2 \times 60 \times 3}{60} + \frac{40 \times 3}{60} = 6 + 2 = 8$$

How would you do the division problem $24 \div 4$ in base 60? Use $\frac{15}{60}$ for $\frac{1}{4}$.[4]

Another skill the Babylonians developed was the extraction of square roots, which was necessary in calculating the areas of geometric figures representing parcels of land. For example, to find the square root of 29, which can be seen to be between 5 and 6, they would first divide 29 by the smaller of the two values:

$$29 \div 5 = 5.8$$

---

[4] $24 \times ,15 = ,360 = \frac{360}{60} = 6.$

Now this result would be too large, so by averaging it with the divisor, they would produce a first approximation:

$$(5.8 + 5) \div 2 = 5.4$$

For further accuracy, you then divide 29 by the result, 5.4, and again average the divisor and the quotient to produce a second approximation as follows:

$$29 \div 5.4 = 5.370 \rightarrow (5.4 + 5.37) \div 2 = 5.39$$

The square root can be obtained to any degree of accuracy by continuing in this way.

Using this square root method, what would the square root of 39 be to two decimal places?[5]

Of course, today a square root can be found in an instant to 10 or more decimal places on a calculator or computer; however, occasionally using one's calculating skills can be a beneficial mental exercise. The Babylonians remarkably calculated the square root of 2 to three sexagesimal fractional digits. This is equivalent to six decimal digits, making the square root accurate to one millionth of its value:

$$\sqrt{2} = 1{,}24{,}51{,}10 = 1 + \frac{24}{60} + \frac{51}{3600} + \frac{10}{216{,}000} = 1.414213$$

This required many steps, but the Babylonians were prodigious calculators and not dissuaded by tedious computation. They produced a very large number of arithmetic tables. Using these tables, they were able to solve equations by various methods, including applying algebraic algorithms. Their approach, in some cases, required a method of iteration that produced more accurate solutions to a problem. Such procedures are still used today to solve problems when a direct method may be very tedious. For example, they were able to

---

[5] $39/6 = 6.5 \rightarrow (6 + 6.5)/2 = 6.25 \rightarrow 39/6.25 = 6.24 \rightarrow (6.24 + 6.25)/2 = 6.245.$

solve certain quadratic equations. The form of the general quadratic equation is:

$$ax^2 + bx + c = 0, \ a \neq 0$$

where a, b, and c are constants. The quadratic formula used to solve this general equation is:

$$x = \frac{-b \pm \sqrt{b^2 - 4ac}}{2a}$$

The Babylonians considered simpler quadratic equations of the form:

$$x^2 + bx = c$$

where a = 1 and c was always positive. Although a quadratic equation could have a negative solution, they only considered a positive solution, since it was applied to a practical problem, such as the dimensions of a rectangular parcel. Their considerations led to a simpler form of the quadratic formula:

$$x = \frac{\sqrt{b^2 + 4c} - b}{2}$$

Applying this simpler form to the quadratic equation:

$$x^2 + x = 5$$

The positive solution is:

$$x = \frac{\sqrt{1^2 + 4(5)} - 1}{2} = \frac{\sqrt{20} - 1}{2} = \frac{4.47 - 1}{2} = 1.74$$

To calculate dimensions of solids, the Babylonians solved certain cubic equations by constructing tables of $n^3 + n^2$, where n is a positive whole number. We present here an elementary example to show clearly how this table was used to solve a cubic equation of the type:

$$ax^3 + bx^2 = c$$

We let a = 1, b = 2, and c = 16:

$$x^3 + 2x^2 = 16$$

If we divide this equation by $2^3 = 8$, it becomes:

$$\frac{x^3}{8} + \frac{x^2}{4} = \frac{16}{8} = 2$$

If we now let $x = 2y$, the equation becomes

$$\frac{(2y)^3}{8} + \frac{(2y)^2}{4} = 2$$

This simplifies to:

$$\frac{8y^3}{8} + \frac{4y^2}{4} = 2 \rightarrow y^3 + y^2 = 2$$

The Babylonians would then consult their table of values of $n^3 + n^2$ and find the closest value of n that satisfies this equation. That value is equal to 1, which is y, and then the solution for x is $x = 2y = 2(1) = 2$.

A team of researchers at the University of New South Wales in Australia has recently revealed a major discovery on one of the Babylonian cuneiform tablets, Plimpton 322, which is 3700 years old. The team, headed by Dr. Daniel Mansfield, found evidence that the Babylonians made significant advances in the field of trigonometry 1500 years before the Greeks, who were great geometers. The tablet contains tables of ratios relating to right triangles, which differ from both the Greek and our present approach, which uses angles. It is a brilliant work that may have advantages over our procedure.

It provides concrete evidence that the Babylonians were well aware of what the Pythagorean theorem says about a right triangle with sides a and b and hypotenuse c: $a^2 + b^2 = c^2$. This predates by more than 1000 years the Greek Hipparchus, who has been regarded as the father of trigonometry.

We list here a summary of some of the impressive mathematical accomplishments of the Babylonians some of which are still in use today:

1. They were able to work with most fractions except possibly those that were indeterminate in base 60, such as $\frac{1}{7}$, which is not a finite decimal in either base 10 or base 60.

    In base 10: $\frac{1}{7} = \frac{1.0}{7} = 0.142857142857...$ (repeats).
    In base 60: $\frac{1}{7} = \frac{8}{60} + \frac{34}{3600} + \frac{15}{216,000} + \cdots = ,8,34,15...$

2. They had a method for extracting square roots.
3. They could solve linear and many quadratic equations.
4. Using tables, they solved some cubic equations.
5. They were quite advanced in their knowledge of the geometry of simple figures, including the circle and the right triangle.

The Chinese

For many years, not much was known about ancient Chinese mathematics. The country was isolated from other civilizations by mountains and seas, so few ideas found their way into the Western world. However, today we know how the development of Chinese mathematics progressed, and that many concepts were known earlier than similar concepts in the Western world. In particular, the famous Pythagorean theorem was known by the Chinese as the *Gougu Rule*, well before the Greek civilization produced its proof. The name of the theorem can therefore be misleading, and credit should be given to the Chinese. Most of their cultural advances date from 1000 BCE onwards, but the progress was not uniform due to constant changes in dynasties. Among the accomplishments of the early Chinese are the development of two number systems, base 2 and base 10, which included fractions and, for the first time, negative numbers. Advances were also made in algebra, geometry, trigonometry, and the theory of numbers.

The deductive geometry of the Greeks, taught widely today in secondary school, relies upon certain accepted axioms, or principles, which provide the foundation for developing more concepts and proving theorems. It is the way mathematics was conceived in the Western world and is practiced today. However, early Chinese mathematics developed very differently, but was no less effective than the mathematics of the Western world. It was primarily learned by solving real-life practical problems involving land measurement, the calendar, architecture, taxes, and commerce. It is fascinating to investigate their methods, and one must marvel at them, as they served the needs of the people very well and certainly have relevance today. A major archeological dig in 1899 uncovered the symbols used by the Chinese number system dating back to 1400 BCE. They are shown in Figure 1.8.

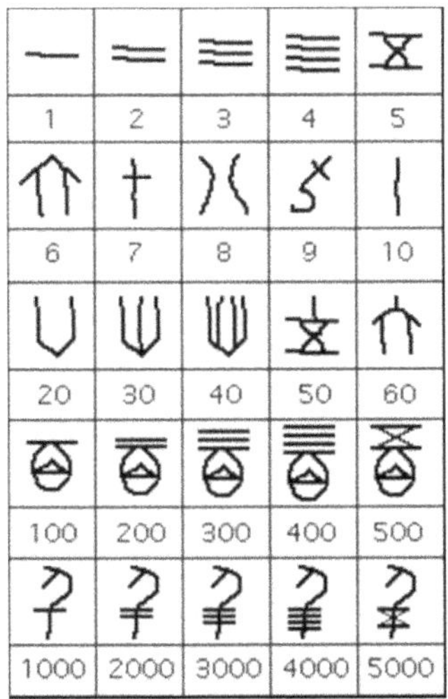

Figure 1.8   Early Chinese numerals.

The evidence that their number system was based on the decimal system is shown on counting boards that were used for calculating as early as 400 BCE (Figure 1.9). A counting board resembles a checkerboard, and the numbers were represented in the squares using bamboo or ivory rods. The place values in the squares moving from right to left were 1, 10, 100, etc., like our present decimal system. A blank square represented zero, so no symbol for zero needed to exist. To multiply or divide by any multiple of ten, the rods could be moved from right to left or vice versa. Chinese counting boards that used rods were unique and not found in any other civilization. They were the precursor of the abacus, which the Chinese developed in the second century BCE.

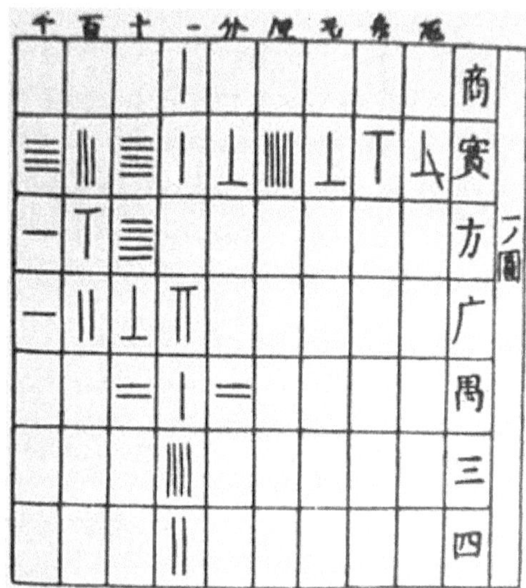

Figure 1.9   Chinese counting board.[6]

The ancient Chinese civilization was ruled by several dynasties, some promoting progress and others repressing it. Two dynasties that saw the origin and development of Chinese mathematics were the Shang (1600–1050 BCE) and the Zhou (1050–256 BCE). The Zhou

---

[6] Public Domain (1996).

was partitioned into several smaller dynasties, one being the Qin (221–207 BCE), which unified China and became an imperial dynasty. The Shang period produced one of the oldest surviving mathematical works, which contains elements of binary (base 2) numbers. It was during the Zhou Dynasty that the Chinese developed a decimal system, arithmetic, algebra, negative numbers, the solution of equations, and some algebraic geometry. Astronomy was the driving force for many of these developments, while others stem from the solution of practical problems. The Qin dynasty became a well-organized and strong military state, effectively applying its knowledge of mathematics and science to practical problems. Its bureaucracy set the tone for Chinese culture for many centuries.

The most outstanding book of Chinese mathematics is the *Jiuzhang Suanshu*, also known as *Nine Chapters on the Mathematical Art*. Its origin is not clear, but it was written possibly sometime around 100 BCE, and its contents seem to have been developed over a long period. However, it remained a major mathematical text and was used for over 1500 years. As mentioned above, the Chinese approach was not deductive but mostly inductive, where the learning was by inference and trial and error. After gaining knowledge by applying a procedure to several cases or continually observing a relationship, a student would master a method of calculation and a conceptual understanding of the theory. The text contains many practical problems described in the *Mac Tutor History of Mathematics* compiled by the University of St. Andrews, Scotland. We present here some of the problems that illustrate the intriguing methods of the ancient Chinese.

The following is from Chapter 6 of the text, which deals with ratio and proportion:

Two runners start from the same position. The faster runner can cover 100 paces in the same time that the slower runner covers 60 paces. The slower runner starts first and covers 100 paces before the faster runner leaves to catch up. How many paces does the faster runner take to catch the slower runner?

This problem can be solved directly today by an algebraic equation. The Chinese might have used a method like the following.

The ratio of paces of the faster runner to the slower runner is 100 to 60, which is the same ratio as 20 to 12, or 10 to 6. Applying these ratios, we can construct a table of the number of paces each runner covers from the start (Figure 1.10). Note in the table, we first add 100 paces twice to the faster runner, while we add 60 paces twice to the slower runner. As the distance between them narrows, we then add 20 paces twice to the faster runner, while we add 12 paces twice to the slower runner. Finally, by only adding 10 paces to the faster runner and 6 paces to the slower runner, we see that the faster runner catches up after 250 paces.

Faster Runner	Slower Runner
0	100
100	160
200	220
220	232
240	244
250	250

Figure 1.10   Chinese runner problem.

To show how this problem is solved using today's algebraic methods, we let $n =$ the number of 100 paces the faster runner must take to catch the slower runner. Then for every $100n$ paces the faster runner moves, the slower runner moves $60n$, where n equals some multiple, not necessarily a whole number. The equation is then:

$$100n = 100 + 60n$$

The left side represents the distance traveled by the faster runner, and the right side the distance traveled by the slower runner. The solution is obtained by subtracting $60n$ from both sides and dividing by 40:

$$40n = 100$$

$$n = \frac{100}{40} = 2.5$$

The faster runner, therefore, travels $100 \times 2.5 = 250$ paces.

Another interesting problem involves what is called the *method of false position,* where trial and error come into play. Variations of this method were used not just by the Chinese but also by the Egyptians, though no link between the two has been found. Consider this simple linear equation:

$$2x + \frac{1}{4}x = 15$$

We first venture a guess, say $x = 8$. Then $2(8) + \frac{1}{4}(8) = 16 + 2 = 18$. The result is greater than 15, so the solution must be smaller than 8. Since we are dealing with a linear equation where results are proportional, we can decrease the guess by a proportional amount and arrive at the correct answer. That amount is the ratio of the number we are looking for, 15, by the number we found, 18. We therefore multiply our guess by this ratio to obtain the correct solution:

$$\frac{15}{18}(8) = \frac{5}{6}(8) = \frac{40}{6} = \frac{20}{3}$$

We urge the reader to check that this is the correct answer.

The method of false position is used for many different types of problems and leads to a clearer understanding of the concepts and better educated guesses to the solution. This is a method of approximation that is sometimes used today to solve difficult problems that cannot be solved directly.

The Abacus

The abacus could be considered the world's first computer. Although its origin is not entirely clear, early forms may have been developed by the Babylonians. However, as it evolved, the type of abacus that became most common was invented by the Chinese around the second century BCE. It remained in use in some less developed parts of the world up until the twentieth century. It utilizes a base 2 and a base 5 system, which together result in a base 10 system. In Figure 1.11, each of the 5 lower beads represents one unit, and each of the 2 upper beads represents five units. By moving the beads up and down, one can add and subtract numbers. Each time five of the lower beads are moved up, an upper bead is moved up, and the lower beads are then moved back down.

Figure 1.11   Chinese-style abacus.[7]

When both upper beads are moved up in one column, a lower bead in the next column is then moved up, and the two upper beads are moved back down. As one becomes proficient in manipulating the beads, addition or subtraction can proceed at a rapid rate.

Multiplication and division can also be done on the abacus, but it is more challenging.

The Egyptians

To fully comprehend the Egyptian civilization (Figure 1.12) and its accomplishments is quite daunting. One can only marvel at what they have achieved, as it has defied complete comprehension due to a lack of sufficient evidence. As early as 3000 BCE, a single Egyptian nation developed with a single ruler. This was a result of the favorable climate, the fertile land caused by the periodic flooding of the Nile, and the surrounding desert, which sheltered it from hostile forces. Egypt, therefore, experienced long periods of peace, allowing its people to achieve a high level of civilization.

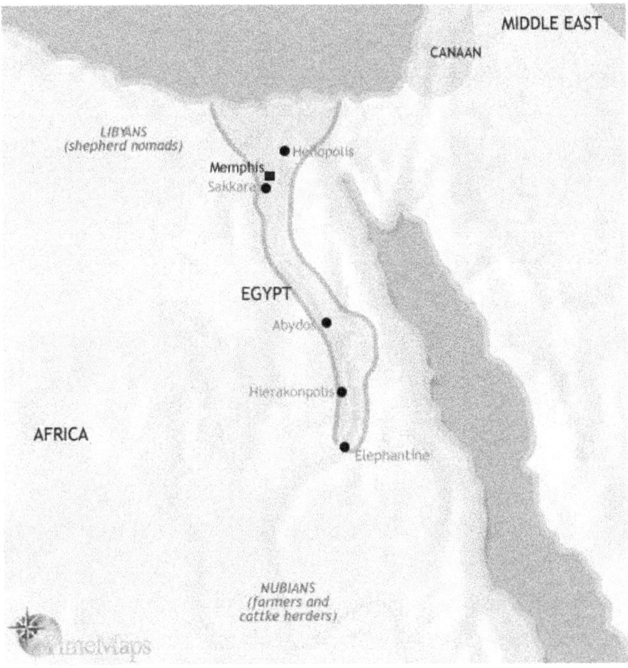

Figure 1.12   The Egyptian civilization.

One of the early scientific achievements was the knowledge of astronomy and the calendar, which was necessary to predict when the Nile would flood. During the period from 1550 BCE to 1070 BCE, the so-called New Kingdom, the Egyptians implemented a 24-hour day. It was based on 24 stars, where 12 of them marked the passage of the night. Since the daylight varied with the seasons, the summer hours were longer than the winter hours. Not until the 14th Century, with efforts made by the Greeks and the Muslims, were the length of the hours made equal in Europe. It is interesting that seemingly separate civilizations established 24-hour days, and that they became the world's standard.

As Egyptian society grew more complex, the need for writing to keep records led to its development, and an arithmetic system evolved to perform computations. Egyptian writing consisted of hieroglyphics, which are found on their stoneware and their amazing temples and pyramids. Hieroglyphs are pictures representing words; however, there is little evidence of how their mathematical calculations were accomplished with their number system. Much of this information, unfortunately, has been lost, as it was written on dried papyrus which has perished. Some did manage to survive, but only because of the dry climatic conditions of the country. The two major mathematical papyri that have been discovered are the Rhind or Ahmes Papyrus (Figure 1.13) and the Moscow Papyrus (Figure 1.15).

A. Henry Rhind (1833–1863), a young Scottish Egyptologist, purchased the papyrus in Luxor in 1858. However, it was produced by the scribe Ahmes, around 1650 BCE, who stated he was copying a document that was written 200 years earlier. The Egyptians developed their hieroglyphic writing around 3000 BCE, marking the beginning of the Old Kingdom period, during which the Pyramids were built. These enormous structures are such an astounding architectural and engineering feat that we have only speculation as to how they were built. Not only was the Egyptian civilization very advanced for its time, but what they were able to accomplish, given their level of technology, shows a remarkable level of technical and scientific ability. An ability that one can argue is equal to, or possibly exceeds, that demonstrated by many other civilizations.

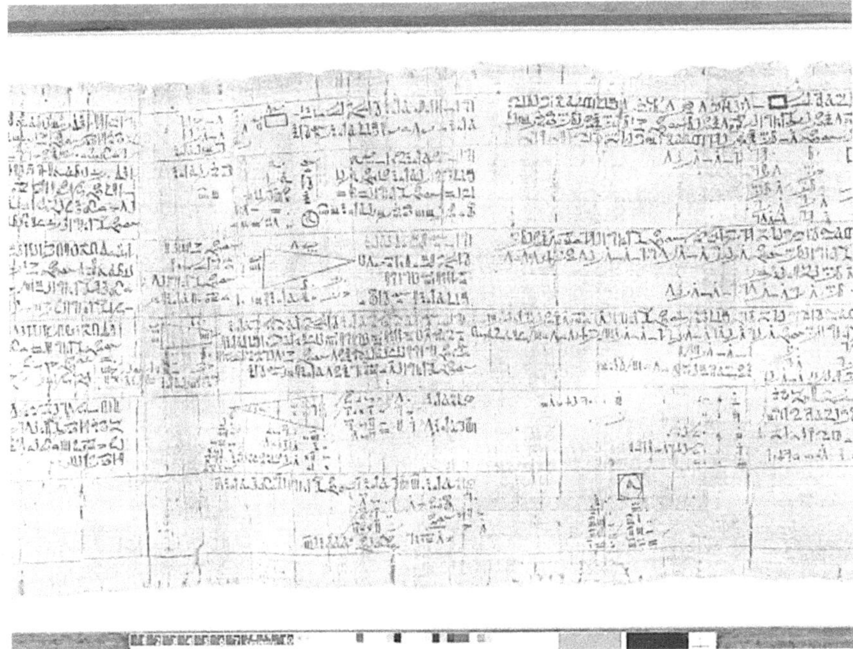

Figure 1.13   Rhind or Ahmes Papyrus.[8]

The Egyptian number system was based on 10 with individual symbols for one, ten, one hundred, one thousand, ten thousand, and one million (Figure 1.14). A vertical staff represents one, a horseshoe shape 10, a rope coil 100, a lotus plant 1000, a pointing finger 10,000, a frog 100,000, and an amazed figure of a god 1,000,000.

1	10	100	1000	10000	100000	1000000

Figure 1.14   Egyptian numerals.

Numbers were written from left to right as we do today, with no more than three symbols on one line. The number 9 was then written as follows:

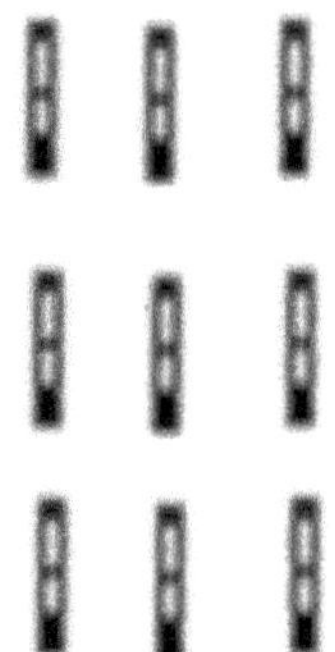

The number 1223 would look like:

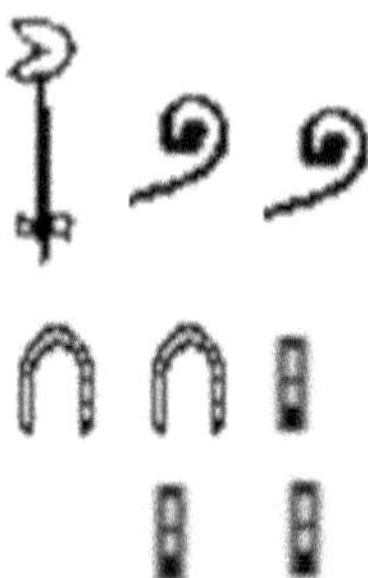

The Egyptian number system was not easy to work with for arithmetical calculations. There was no zero because it was not needed to represent any number. Despite this weakness, they devised ingenious ways to multiply and divide numbers and work with fractions, which became necessary as their trade and economic system developed. They cleverly employed only addition to multiply and divide numbers. The Ahmes and the Moscow papyrus (Figure 1.15) provide almost all the evidence we have of Egyptian mathematics. This system employed a hieratic script and a cursive form of writing that was only used traditionally for religious texts and literature.

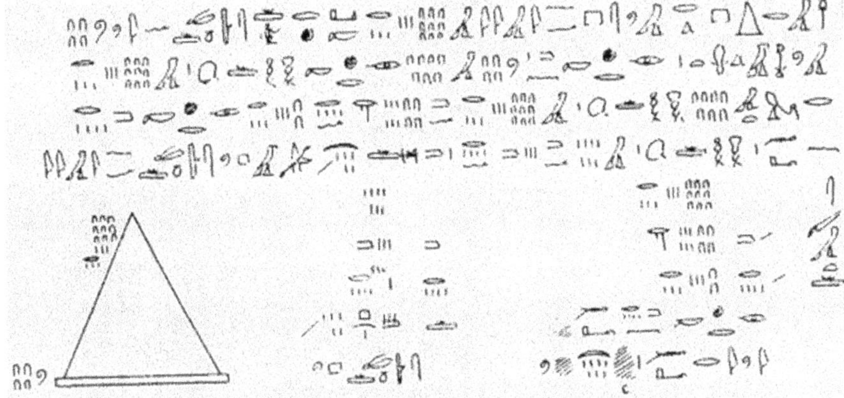

Figure 1.15   Moscow Papyrus problem.

There are eighty-seven problems in the Ahmes Papyrus and twenty-five in the Moscow Papyrus. Seven of the twenty-five problems are geometry problems and include finding areas of triangles and determining the surface area of a hemisphere. The problems are mostly practical applications where numbers were not thought of as abstract ideas but as sort of concrete quantities.

Here is an example of problem 24 in the Ahmes Papyrus:

A number added to a quarter of number is 15. What is the number?

The Egyptians used the method of false position, as did the Chinese, to solve such a problem. The solution involves taking an educated guess at the answer and then correcting any errors to arrive at the answer. Ahmes first guesses 4 as the answer and calculates the result of taking 4 and adding one-quarter of the number 4 to itself:

$$4 + \left(\frac{1}{4}\right)(4) = 4 + 1 = 5$$

The answer above, 5, is not correct, but when you compare it to the desired result, 15, you see it needs to be three times greater. Therefore, applying that reasoning, Ahmes multiplies 4 by 3 to arrive at the correct answer 12. So, by comparing the result of one's guess and then applying a proportional correction factor, one arrives at the correct answer of 12. This method of false position existed for almost three thousand years. Egyptian mathematicians applying this method would most likely develop a keen sense of numbers and soon would be able to make more accurate guesses at the answer to a problem.

The Egyptians employed only unit fractions in their arithmetic, and for some reason, they included only one fraction that was not a unit fraction: $\frac{2}{3}$. Here is a set of problems from the Ahmes Papyrus that, while less practical, involves the fair division of loaves of bread:

> *What is the fairest way to divide n loaves of bread among 10 people where n equals the following numbers: 1, 2, 6, 7, 8, and 9?*

The solutions to these fair division problems can be expressed using only unit fractions as follows:

n loaves	Loaves each man receives
1	$\frac{1}{10}$ loaf
2	$\frac{2}{10} = \frac{1}{5}$ loaf
6	$\frac{6}{10} = \frac{1}{2} + \frac{1}{10}$ loaf
7	$\frac{7}{10} = \frac{1}{2} + \frac{1}{5}$ loaf
8	$\frac{8}{10} = \frac{1}{2} + \frac{1}{5} + \frac{1}{10}$ loaf
9	$\frac{9}{10} = \frac{1}{2} + \frac{1}{5} + \frac{1}{5}$ loaf

Many other fractions can be expressed as the sum of unit fractions, such as:

$$\frac{5}{8} = \frac{1}{2} + \frac{1}{8} \quad \text{and} \quad \frac{11}{12} = \frac{1}{2} + \frac{1}{3} + \frac{1}{12}$$

It can also be shown that all unit fractions between 0 and 1 can be expressed as a sum of different unit fractions. For example:

$$\frac{1}{5}=\frac{1}{6}+\frac{1}{30} \quad \frac{1}{7}=\frac{1}{8}+\frac{1}{56} \quad \frac{1}{9}=\frac{1}{10}+\frac{1}{90}$$

The concept of fair division, as illustrated in the above table, was significant in Egyptian culture and is reflected in various problems. In fact, even though Egyptian society was class-structured, social mobility was not impossible. Peasant boys who mastered reading and writing could become scribes and work their way up to high ranks in the government.

A fascinating aspect of Egyptian mathematics is the way they performed multiplication. It was based on powers of 2, or the binary system. They compiled large tables of powers of 2, and the Egyptians knew that any number could be made up as the sum of unique powers of 2.

Consider the multiplication problem 23×9. The larger number 23 is first broken down into powers of 2:

The largest power of 2 less than or equal to 23 is $16 = 2^4$.
Then $23 - 16 = 7$
The largest power of 2 less than or equal to 7 is $4 = 2^2$.
Then $7 - 4 = 3$
The largest power of 2 less than or equal to 3 is $2 = 2^1$.
Then $3 - 2 = 1$
The largest power of 2 less than or equal to 1 is $1 = 2^0$.
Then $= 1 - 1 = 0$

Therefore $23 = 16 + 4 + 2 + 1$. Then, taking the smaller number 9 and continuously doubling it, multiples of 9 are obtained that are products of the powers of 2 as follows:

1   $9 = 1 \times 9$
2   $18 = 2 \times 9$

$$4 \quad 36 = 4 \times 9$$
$$8 \quad 72 = 8 \times 9$$
$$16 \quad 144 = 16 \times 9$$

One stops at the largest product that is smaller than 23×9, which is 16×9. Then, since $23 = 16 + 4 + 2 + 1$, one calculates the sum of the four products: $16 \times 9 = 144$, $4 \times 9 = 36$, $2 \times 9 = 18$, and $1 \times 9 = 9$. The answer will equal the product 23×9:

$$23 \times 9 = 144 + 36 + 18 + 9 = 207$$

This type of multiplication may seem tedious compared with our present methods. However, an Egyptian familiar with powers of two, as we are with multiplication tables, would be able to readily apply this method to multiply numbers. When one considers that the Egyptians had to calculate with their hieroglyphic form of writing numbers, one cannot help but be impressed with the mathematics this civilization developed over four thousand years ago.

The Elamites

As archeological digs continue, more and more evidence of ancient civilizations is revealed, providing continuous amazement at how old and advanced these civilizations were. The Elamite civilization flourished at the same time as the Babylonians and was located along the shores of the Persian Gulf. (See Figure 1.16.) They were among the leading political forces of the Ancient Near East, known today as the Middle East. Presently, we have limited knowledge of their history, which is divided into three periods based on the dynasties and the influence of neighboring civilizations:

> **Old Elamite period:** c. 2700 – c. 1500 BCE (earliest documents until the Sukkalmah Dynasty)
> **Middle Elamite period:** c. 1500 – c. 1100 BCE (Anzanite dynasty until the Babylonian invasion of Susa)

Neo-Elamite period: c. 1100 – 540 BCE (characterized by Assyrian and Median influence; 539 BC marks the beginning of the Achaemenid period)

Figure 1.16   Elamite civilization.[9]

The Elamites developed an advanced civilization, similar to that of the Babylonians, and were accomplished in their art and architecture. Recently, their script has been deciphered from clay tablets found in southwest Iran. These tablets reveal significant geometric knowledge, including the Pythagorean theorem, the geometry of circular figures, and the volumes and areas of various geometric shapes.[10] The Elamite

---

[9] Morningstar1814 via Wikimedia Commons (CC-BY-SA).

[10] https://www.researchgate.net/profile/Nasser-Heydari-2.

civilization interacted with the Babylonians, and though evidence is limited, it appears that their science and mathematics were similar to those of the Babylonians. The Elamites may have even developed some of it.

The Maya

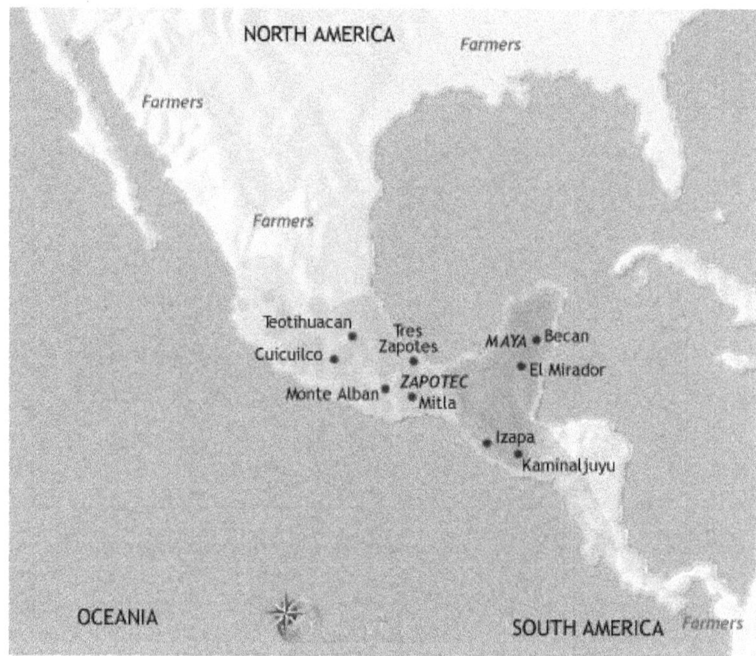

Figure 1.17   The Maya civilization.

The ancient civilizations of Central America (Figure 1.17) must also be included here, as we now know their societies were extraordinarily complex and they were masters of mathematics, science, and astronomy. As recently as December 2022, more than 1000 Maya settlements in northern Guatemala have been located, some dating earlier than previously unearthed Maya sites and going back to 1000 BCE.[11] Their large, sophisticated architectural structures are

---

[11] https://www.livescience.com/lidar-maya-civilization-guatemala.

accurately aligned to the movements of the Sun, Moon, and planets. This allowed them to predict astronomical events precisely and to develop a very exact calendar system. These accomplishments required an advanced mathematical system, which the Maya no doubt possessed. We have yet to unearth evidence of the origin of Maya mathematics, but when the system was developed, it most likely contained concepts which were more advanced than any other in the world at the time. Their sophisticated number system was a base 20, or vigesimal, system, most likely stemming from the use of both fingers and toes to calculate. See Figure 1.18.

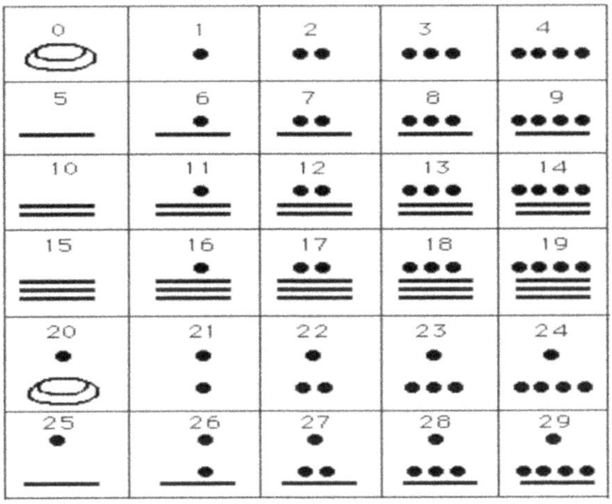

0	1	2	3	4
5	6	7	8	9
10	11	12	13	14
15	16	17	18	19
20	21	22	23	24
25	26	27	28	29

Figure 1.18   Maya positional number system.

Note that the Maya number system contains only three symbols: a dot, a bar, and two concentric circles, which are used for zero and multiples of 20. The Maya number five is considered important and is assigned a bar to represent it, like a bead on an abacus. This may stem from the number of fingers on one hand or the number of toes on one foot. The position of the dots and the bars signifies their value. Each dot on the lowest or first line equals 1, and each bar equals 5. Each dot on the second line above the first equals 20, and each bar equals 5 × 20. On the third upper line, each dot equals 18 × 20 = 360, and each bar equals $5 \times 18 \times 20 = 1800$. The Maya chose 18×20 for each dot on the third line because 360 is close to the number of days in the year, and the numbers were tied to the calendar and astronomical calculations. The place value of the fourth upper line was 18×20^2, the fifth upper line 18×20^3, etc. For example, consider the number 37,583 in the Maya system:

$$37{,}583 = (5 \times 18 \times 20^2) + (4 \times 18 \times 20) + (7 \times 20) + (3 \times 1)$$

It would be written as follows, with the numbers **3, 7, 4, 5** from bottom to top:

This positional nature of the Maya number system is not a true positional system like our decimal system. A true base 20 system would have the place values: 1, 20, $20^2 = 400$, $20^3 = 8000$, and so on, like our decimal system, where the columns from right to left are 1, 10, $10^2 = 100$, $10^3 = 1000$, etc. Remarkably, the Maya were one of the earliest civilizations to invent a unique symbol for the number zero, which speaks to their mathematical sophistication. This is in contrast to the early Babylonians, who did not have a unique number for zero. To distinguish the number 6 from the number 60, they used a space and later on devised a wedge-like symbol for a zero between figures. The Romans and the Greeks had no symbol for zero. The two concepts of zero as a placeholder indicating no value, and as a real number in our decimal system, did not fully arise in the Western world until the 7th century CE in India. It is what makes our decimal system so remarkably easy to use.[12] We owe credit to the Maya for first

---

[12] See "History of Zero" in *Mathematics: Its Historical Aspects, Wonders and Beyond,* Posamentier and Kramer, World Scientific, 2022.

developing a symbol for zero before it arose in the Western world. There is no evidence that the Maya developed methods of multiplication or division or worked with fractions, yet such operations were possible with their number system. However, with only sticks placed at right angles, they were able to view astronomical objects and make measurements with incredible accuracy. They calculated the length of the year to be 365.242 days (the present value is 365.242198 days) and the lunar month to be 29.5308 days (the present value is 29.53059 days). The Maya were one of the great ancient and sophisticated civilizations of our world, which were very advanced mathematically.

Though thousands of years have passed since the dawn of mathematics, many early civilizations were very adept at mastering the practical mathematics that most people need and use today. Some went further in their mastery, convincing us that they possessed skills and abilities no less capable than any we experience today.

Chapter 2

The Ancient Greeks and Romans

The ancient Greek civilization flourished for almost seven hundred years from about 800 BCE to its domination by the Roman Empire in 146 BCE. They made great achievements in philosophy, art, architecture, astronomy, science, medicine, literature, and dramatic theater. In addition, the Greeks were the most prolific mathematical culture of the ancient world, whose impact is still very much alive today. The deductive reasoning formulated by their culture provides a basis for many of our present mathematical concepts. The most famous work of the Greeks, still taught in schools today, is Euclid's *Elements*, which consists of thirteen books covering plane and three-dimensional geometry, number theory, and irrational quantities. It is the oldest significant treatise that applies deductive logic to a large body of mathematics and is truly a mathematical masterpiece.

Euclid of Alexandria lived from 365 BCE to 265 BCE and compiled his treatise on geometry around 300 BCE. However, the oldest extant copy dates to 888 CE, as all previous copies either perished or were destroyed. It is a compilation of mathematical proofs and ideas from many mathematicians, and some, perhaps, from Euclid himself. Anybody who has taken high school geometry has seen some of these concepts and proofs. He was an exceptional mathematics teacher who taught at Alexandria in Egypt. Unfortunately, little is known about the rest of his life except that he lived during the time of the first Ptolemy, was a pupil of Plato, and was older than Archimedes.

The Greeks

Before we reveal more about Euclid's *Elements*, let us look at the significant mathematical developments of the Greek civilization that preceded and informed Euclid's work. Ancient Greece consisted of many independent islands and city-states, which had their own currency and measurement standards. Their practical number systems, therefore, varied somewhat as they were mainly used for money transfers and other business transactions. The number systems were not positional like the Babylonians, but were like the Egyptian and the Roman systems, which used an additive principle. Figure 2.1 is one example of these so-called acrophonic numbers.

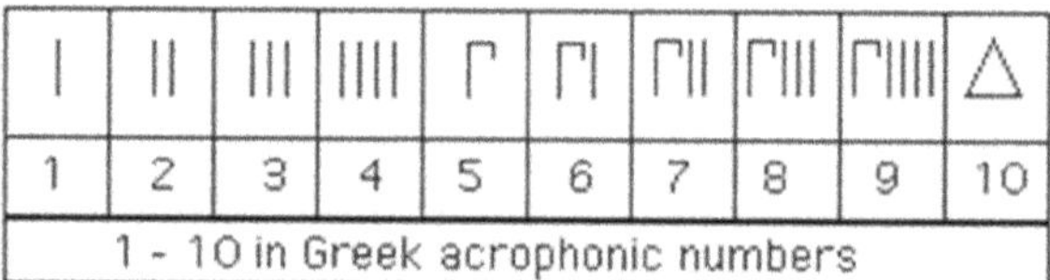

Figure 2.1   Greek practical numbers.

The numbers in Figure 2.1 have 10 as a base and 5 as a secondary base. Though the Greeks were influenced by a base of ten, the Greek system of money was not entirely decimal. The unit of money was the drachma, which was divided into smaller units, where $\frac{1}{6}$ of a drachma was called an obol and 100 drachmas equaled a mina. By 400 BCE, the acrophonic system was replaced with an alphabetic system where the numbers 1 to 9 were assigned separate letters, as were the numbers 10 to 90 and the numbers 100 to 900 (Figure 2.2). The Greek alphabet had only 24 letters, so three obsolete letters were employed: fau for 6, koppa for 90, and sampi for 900.

To not confuse numbers with letters, a keraia sign, which is like an apostrophe, was written after the letters. For example, the number 539 was written as ΦΛΘ' (500 + 30 + 9). The Greek alphabetic numbers were limited in that the largest number that could be written was 999 when using three digits. The Greeks overcame this problem by placing the keraia sign to the left of the letters for 1 to 9. This meant they were multiplied by 1000 and therefore represented the

A	B	Γ	Δ	E	Ϛ	Z	H	Θ
α	β	γ	δ	ε	ϛ	ζ	η	θ
1	2	3	4	5	6	7	8	9

I	K	Λ	M	N	Ξ	O	Π	Ϙ
ι	κ	λ	μ	ν	ξ	o	π	ϙ
10	20	30	40	50	60	70	80	90

P	Σ	T	Y	Φ	X	Ψ	Ω	ϡ
ρ	σ	τ	υ	ϕ	χ	ψ	ω	ϡ
100	200	300	400	500	600	700	800	900

Figure 2.2   Greek alphabetic numbers.

numbers from 1000 to 9000. For example, the number 9539 was written 'Ϟ ΦΛΘ'. For everyday use, these symbols sufficed. However, some of the mathematicians created other symbols to represent very large numbers for their research.

Archimedes of Syracuse (287–212 BCE) was the greatest mathematician of ancient Greece and is considered by some to be one of the greatest mathematicians of all time. (See Figure 2.3.) His outstanding achievements are almost too numerous to list and are particularly impressive given the lack of technology that existed in Archimedes' time. His contributions in geometry revolutionized the subject, and

Figure 2.3   Archimedes of Syracuse.

his methods were a forerunner of integral calculus, which was developed a thousand years later. In his early years, he achieved fame by producing many mechanical inventions, some of which were used as engines of war. However, he felt strongly that pure mathematics, not used for political or personal gain, was the only worthy pursuit.

He considered scientific and engineering discoveries that were used for mere advancement and profit to be morally distasteful, and he preferred to devote himself to subjects that possessed beauty and elegance and were not associated with the more sordid aspects of life. He wrote very little about his mechanical and engineering contributions, preferring to elaborate on his mathematical ideas. Let us look at some of his most impressive mathematical contributions, some of which did not fully materialize until many years later.

1. *Approximation of the Number Pi (π)*

The number designated by π is the ratio of the circumference of a circle to its diameter. Its exact value cannot be written as a fraction. In school, it is approximated by the fraction $\frac{22}{7}$, or the decimal 3.14, which suffices for most calculations. If it were equal to a fraction, it could be written as a decimal with a repeating pattern. For example, $\frac{22}{7} = 3.142857142857142857\ldots$ where the six digits 142857 repeat to infinity. The number π is an irrational number, and its infinite decimal representation has no repeating pattern: $\pi = 3.141592653589\ldots$ The accuracy of π, therefore, challenged mathematicians for hundreds of years until the discovery of calculus in the 17th century, which produced an infinite formula that exactly expresses its value:

$$\pi = 4\left(1 - \frac{1}{3} + \frac{1}{5} - \frac{1}{7} + \frac{1}{9} - \ldots\right)$$

Before Archimedes, π was known to be a little larger than 3. The Rhind Papyrus (ca. 1650 BCE), mentioned in Chapter 1, records that the ancient Egyptians calculated the value 3.1605 for pi. Archimedes brilliantly employed a *method of exhaustion*, devised by the great Greek mathematician Eudoxus of Cnidus, to calculate the value of π between $3\frac{10}{71}$ and $3\frac{1}{7}$, which is between 3.1408 and 3.1429. This was a remarkable result which Archimedes obtained by both circumscribing a circle with a regular polygon and inscribing the same circle with another regular polygon. (See Figure 2.4.)

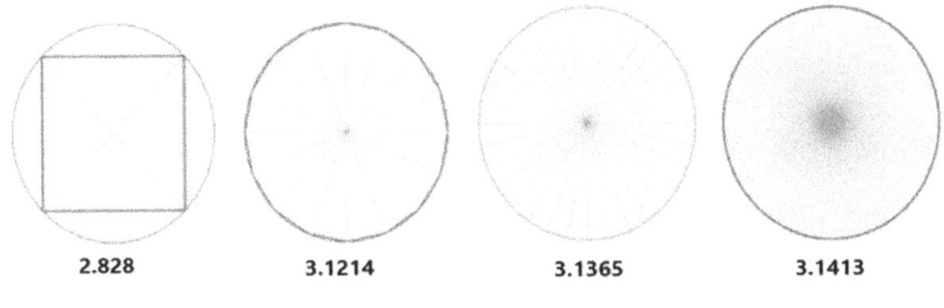

Figure 2.4   Inscribing polygons to approximate π.

By increasing the number of sides of the polygon up to 96, he calculated the perimeters of the inscribed polygon and the circumscribed polygon and obtained approximations for the circumference of the circle. Then, using the formula $\pi = \frac{C}{D}$, where C is the circumference of the circle and D is the diameter of the circle, Archimedes calculated a range of values for π. This was not an easy calculation to do by hand, which involved some advanced trigonometry. As far as we know, he was the first to calculate the value of π to this level of accuracy, and it was an impressive result as π is approximately 3.1419 to four decimal places. Using this method, he was also able to calculate square roots accurately and perfect a method of integration which allowed him to find areas, volumes, and surface areas of many bodies. This integrative method was one of the basic concepts that the discovery of calculus was based on 2000 years later.

2. *Geometrical Contributions*

As mentioned earlier, Archimedes made many contributions to geometry. His books include *On the Sphere and Cylinder, Spirals, Measurement of a Circle*, and *The Sand Reckoner.* In the first book, *On the Sphere and Cylinder*, Archimedes proved two major formulas in geometry: the surface area of a sphere is $4\pi r^2$ and the volume of a sphere is $\frac{4}{3}\pi r^3$, where r is the length of the radius. These formulas are now commonplace in basic geometry. In *On Spirals*, he defines the Archimedean spiral as one that increases the same distance with each revolution. (Figure 2.5.) That is, a line drawn outward from the origin of the

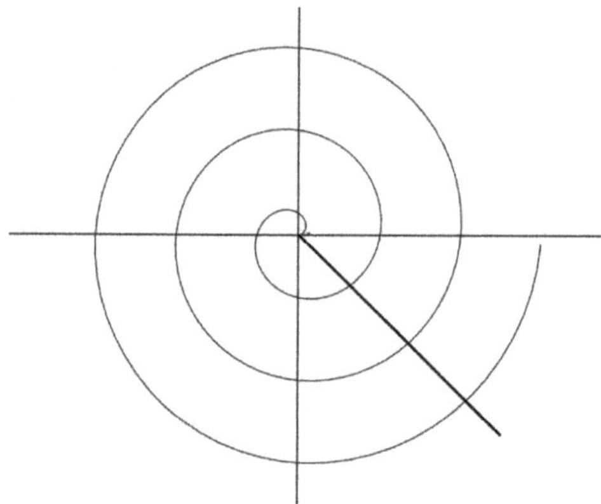

Figure 2.5   Archimedean spiral.

spiral, known as a radius vector, intercepts each turn of the spiral equidistantly. He shows several properties of the spiral, including how to find areas of portions of the spiral. Spirals are found in nature, and years later, in 1525, Albrecht Dürer (1471–1528), a German painter, printmaker, and theorist of the German Renaissance, described another spiral called a "Growth Spiral" which is found on snail shells and other shells. The famous scientist and philosopher René Descartes (1596–1650), more than a century later, studied this spiral, which is a logarithmic spiral that increases in size exponentially. That is, the radius of each turn is twice the radius of the previous turn. The radii, therefore, form a geometric progression (Figure 2.6).

(a)   (b)

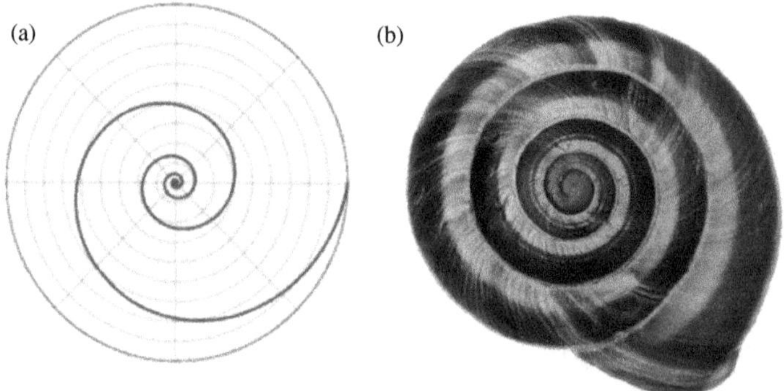

Figure 2.6   (a) Logarithmic spiral and (b) Snail shell.

3. *The Sand Reckoner*

One exceptional work of Archimedes is "The Sand Reckoner," where he proposes a number system capable of expressing numbers as large as 8×10^{63} in modern notation. This number is 8 followed by 63 zeros. He claims that a number this large represents the number of grains of sand that could fill the universe. In this remarkable claim, he also states that the astronomer Aristarchus (310–230 BCE) proposes that the Sun is the center of the solar system and the planets revolve around it. This brilliant and correct Greek discovery took almost 2000 years to be fully accepted by the world until it was proposed by Nicholas Copernicus (1473–1543). Claudius Ptolemy (100–160 CE) proposed a geocentric system, with the Earth as the center, which was not true. However, the Church adopted it, and it dominated their worldview until Copernicus.

Some other outstanding mathematicians of ancient Greece that we cannot omit include the famous Pythagoras of Samos (570–490 BCE). He is considered the first pure mathematician and was an exceptional figure in the history of mathematics. He led a somewhat secret society of Pythagoreans that combined religion and science; unfortunately, so little is known about him. The famous theorem named after him, the Pythagorean theorem, is ubiquitous. However, it was certainly known by many earlier advanced civilizations whose achievements would have required knowledge of the geometry of a right triangle. There does exist an ancient Babylonian tablet, IM 67118, where the theorem is used to calculate the diagonal of a rectangle. This tablet predates Pythagoras' birth by more than a thousand years. The theorem is erroneously named after him only because the Pythagoreans might have been the first to prove it deductively. Anybody who has taken basic geometry may forget most of the concepts, but can almost always recite the Pythagorean theorem in this form: $a^2 + b^2 = c^2$. Here is what some believe to be Pythagoras's proof of the famous theorem: Construct a square with four right triangles having legs a and b and hypotenuse c, as shown in Figure 2.7.

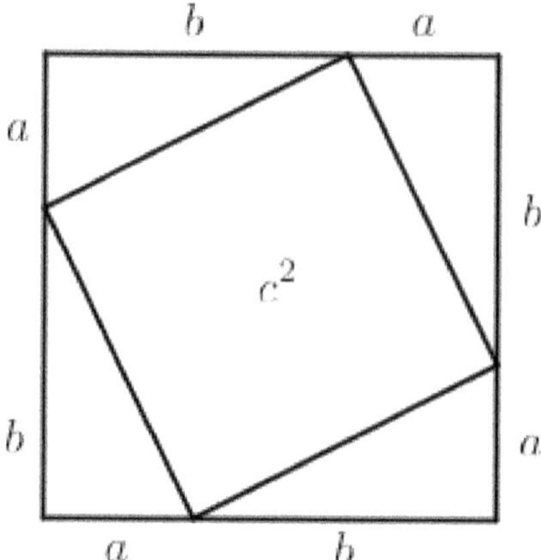

Figure 2.7   Pythagorean theorem 1.

The side of the large square is then $a + b$, and a smaller square is formed in the center whose side is c and whose area is c^2. Now using the same four triangles, an identical square having the same side, $a + b$, can be formed as shown in Figure 2.8.

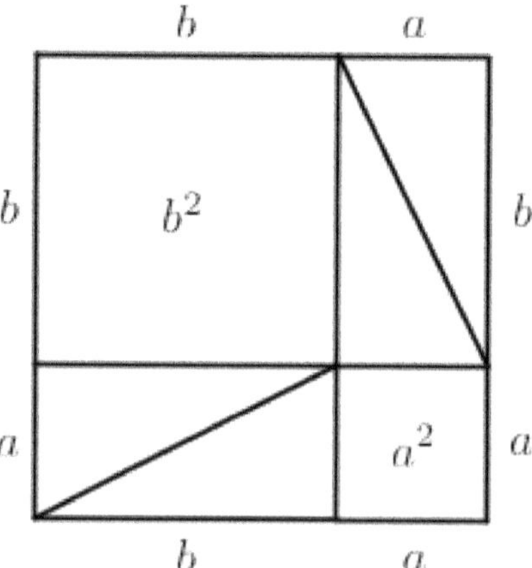

Figure 2.8   Pythagorean theorem 2.

This square contains the four triangles and two smaller squares whose areas are a^2 and b^2.

Now, since the two large squares have the same sides, $a + b$, the areas of both large squares are equal. Therefore, we have the following equality between the areas of the two squares:

$$(\text{Areas of four triangles}) + a^2 + b^2 = (\text{Areas of four triangles}) + c^2$$

Removing the areas of the four triangles from both sides of the equality, we are left with $a^2 + b^2$ equal to c^2, and we have the Pythagorean theorem.

Eudoxus of Cnidus (390–340 BCE) was a prominent ancient Greek mathematician and astronomer, known for his contributions to proportion theory, planetary models, and the method of exhaustion. Eudoxus, along with Plato's treatise *Theaetetus*, greatly influenced Euclid. It therefore appears that Euclid most likely also studied at Plato's Academy. Apollonius of Perga (262–190 BCE) was known as "The Great Geometer." There is scant evidence about his life, but his contributions have had a very great impact on the development of mathematics. His famous book *Conics* introduced terms which are familiar to us today, such as the parabola, ellipse, and hyperbola. Hipparchus of Rhodes (190–120 BCE) was a brilliant astronomer and mathematician who compiled examples of trigonometric tables. He is known as the father of trigonometry. He calculated the length of the year to an amazing accuracy of six and a half minutes. He discovered the precession of the equinoxes, caused by the polar axis of the earth rotating, to within 10% of the present value.

Eratosthenes (276–194 BCE) accomplished an amazing feat by measuring the circumference of the earth to be 24,858 miles! This remarkable result is within 0.2% of the actual figure, which is 24,901 miles.[1] Other great contributions by ancient Greek mathematicians are too numerous to mention here, and the reader is referred to https://mathshistory.st-andrews.ac.uk/, where extensive material can be found.

---

[1] See *Mathematics: Its Historical Aspects, Wonders and Beyond*, pp. 173–175, Posamentier and Kramer, World Scientific, 2022.

Euclid's *Elements*

As mentioned earlier, Euclid's *Elements* consists of thirteen books covering plane and three-dimensional geometry, number theory, and irrational quantities. His plane geometry is based on five assumptions, or postulates, and some definitions. The first three books deal with the existence of points and the construction of straight lines and circles. The fourth postulate states that all right angles are equal, which seems unnecessary except that it warrants that the position and orientation of a right angle does not alter its shape. The fifth, and most famous, postulate states that one and only one line can be drawn parallel to a given line through a point not on the line. This assumption justifies the geometry of flat surfaces, or "plane geometry." For over 2000 years, plane geometry had biblical stature and was accepted as the geometry of the world. Not until the 19th century was it possible for this postulate to be altered and the change accepted, resulting in two types of non-Euclidean geometry. Euclid's postulates and their alteration are discussed in more detail in Chapter 8.

Of the thirteen books that comprise Euclid's *Elements*, books numbered one through six cover a significant amount of plane geometry. The first three books deal with the existence of points and the construction of straight lines and circles, while Book Four examines problems involving circles. Book Five discusses the concept of proportion developed by Eudoxus, which serves as an important basis for much of the geometry. Book Six presents applications of proportionality, while books seven through nine deal with the theory of numbers. Book Ten deals with quantities that are not fractions or rational numbers, like pi and the square root of 2. The Greeks struggled somewhat with the acceptance of these incommensurable quantities and did not consider them the same as other numbers. This is also true about negative numbers. The Greeks were strictly geometers, and an immeasurable distance or a negative distance had little meaning for them in their geometry. They also struggled with the concept of infinity. The Greek philosopher Zeno of Elea (495–430 BCE) is considered the first to deal with the concept of infinity. He appears to have been a self-taught young man from the country who came up with many famous paradoxes, some of which involve infinite series of numbers.[2] Zeno's arguments appear to be the first examples of a method of proof called *reductio ad absurdum* which leads to an absurd result, causing one to question the original premise. One of his most popular paradoxes deals with the impossibility of motion as follows.

Suppose you want to travel between two destinations, A and B, which are a mile apart, and you are traveling at one mile per hour. The premise is that you will arrive at B after one hour. After $\frac{1}{2}$ hour, you will have traveled $\frac{1}{2}$ the distance between A and B. Then, after another $\frac{1}{4}$ hour, you will have traveled $\frac{3}{4}$ of the distance, and after another $\frac{1}{8}$ of an hour, you will have traveled $\frac{7}{8}$ of the distance. Carrying on with this argument and continually halving the remaining distance to be traveled, it appears that you will get closer to point B but never quite get there. (See Figure 2.9.) The paradox lies in the fact that you know it is possible to travel the total distance, but this argument seems to contradict it.

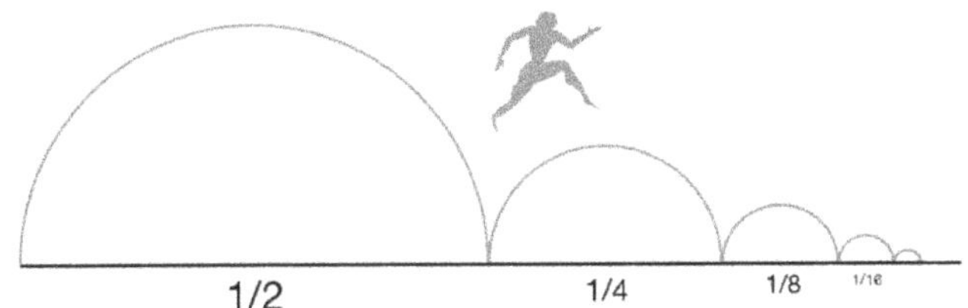

Figure 2.9   Zeno's paradox.

Zeno's paradoxes have intrigued, puzzled, and amused not just mathematicians, but physicists, philosophers, and perhaps many others for over two thousand years. The famous mathematician and philosopher Bertrand Russell (1872–1970) called Zeno's paradoxes "immeasurably subtle and profound."

This type of reasoning, which leads to an absurd result, is similar to an indirect mathematical method of proving something not true by assuming it is true and showing it leads to a contradiction. If one's

---

[2] See *Mathematics: Its Historical Aspects, Wonders and Beyond*, pp. 48–56, Posamentier and Kramer, World Scientific, 2022.

premise is true, then correct logical reasoning should lead to a true result. If the result is incorrect, then the original assumption must be false. Euclid used this method to prove that prime numbers, which are numbers that have no divisors except one and the number itself, are infinite. He assumed the number of primes was finite and then showed that, if that is true, it is possible to construct a new prime that is not one of the original finite ones. Hence, the original premise, that the number of primes is finite, must be false.

To illustrate this type of reasoning, we show here an example of an indirect proof that $\frac{0}{0}$ cannot be equal to 1. First assume that $\frac{0}{0} = 1$. Now consider the accepted true statement that: $2 \times 0 = 1 \times 0$. If we divide both sides by 0, we have $2 \times \frac{0}{0} = 1 \times \frac{0}{0}$. Now if $\frac{0}{0} = 1$, then $2 \times 1 = 1 \times 1$ which means $2 = 1$, which is clearly not true. Hence, we must conclude that $\frac{0}{0} \neq 1$. Mathematicians would rather prove ideas directly as they feel they are stronger arguments. However, they will resort to an indirect proof when a direct proof is very difficult.

The concepts of negative numbers, infinity, and incommensurable numbers, which are numbers that cannot be measured exactly, such as π and $\sqrt{2}$, became more mathematically clear with the discovery of calculus in the 17th century, which will be discussed in Chapter 5. Books eleven through thirteen of Euclid's *Elements* deal with the geometry of three dimensions and contain mainly the work of Eudoxus and the treatise by Theaetetus. They discuss spheres and the five regular polyhedra, including a proof that there are only five. (See Figure 2.10.)

Tetrahedron	Cube	Octahedron	Dodecahedron	Icosahedron
Four faces	Six faces	Eight faces	Twelve faces	Twenty faces

Figure 2.10   Five regular polyhedra (Platonic Solids).

Euclid's *Elements* may lack some rigor, but it is still quite magnificent for its content and clarity, considering the time in which it was produced. In 1756, it was first translated into Latin and English by the Scottish mathematician Robert Simson (1687–1786). Eventually, it became the basis for the secondary school geometry course taught in England and the United States. It is considered by many to be the greatest mathematical textbook ever written.

Classical Greek Problems

Greek mathematics grappled with three classical problems that mathematicians then struggled with throughout the history of mathematics. One can find evidence of them in some of the oldest mathematical manuscripts, up to those in the present day. They are worth looking at because of their keen interest and the fascinating number π. The problems are:

1. Squaring the Circle
2. Doubling the Cube
3. Trisecting an Angle

The Greeks sought solutions to these problems in the following disciplined way. They had to be solved by constructing them geometrically with an unmarked ruler and a compass. Given that the original measurements are rational numbers (whole numbers or fractions), the solutions would also have to be rational numbers so they could be constructed from the original measurements. This is where the difficulty lies. The most famous of these problems is Squaring the Circle, also known as "The Quadrature of the Circle", which means: Given a circle of a certain radius, can one construct a square with the same area? This problem is very intriguing and not solvable geometrically, because the area of a circle with a radius that is a whole number or a fraction, that is, a rational number, will have an area that is *not* a rational number because of the number π in the area formula for a circle: $A = \pi r^2$. If it were possible to construct a square with the same area as the circle, then the number π could be determined as follows. Suppose the side of the square with the same area as the circle equals the rational number 3. Then the area of the square is $3^2 = 9$, and this will equal the area of the circle: $9 = \pi r^2$. By dividing both sides by π and then taking the square root, we have: $r = \sqrt{\frac{9}{\pi}}$. Now π is an irrational number, with an infinite decimal without a pattern among any of its digits: $\pi = 3.1416926535898...$ The radius of the circle would then also be an irrational number and could not be constructed geometrically. Note that today, with our super-computers, the number π has been calculated to over 30 trillion digits, and no patterns among the digits have been found!

The second classical problem states: Given a cube, can you construct another cube whose volume is twice the volume of the given cube? This problem also involves an irrational number. Suppose the side of the first cube is the rational number 2, then its volume is $2^3 = 8$. The volume of the doubled cube is then $2 \times 8 = 16$, and the side of the doubled cube is $\sqrt[3]{16}$. This cube root is not a rational number. When the side of the original cube is a rational number, then the side of the doubled cube will always be an irrational number. Again, we have a quantity that cannot be constructed geometrically, and its measurement, therefore, can also not be determined. The third classical problem presents a similar scenario and cannot be constructed geometrically. This, however, has not prevented people over the years from endlessly claiming to have "solutions" to these three problems, but none have been proven correct. It was not until the 19th century that these problems were clearly shown to be impossible.[3]

The Romans

The Roman civilization, with its many achievements in engineering, architecture, government, medicine, education, irrigation, central heating, and sewage systems, did not produce any significant contributions in mathematics. It is hard to accept, given their advanced civilization, that they seemed uninterested in the logical or theoretical investigation of number systems and geometry that the Greeks explored. The Greek civilization had developed all the mathematics that they needed before and during Roman times. The creation of the Roman calendar instituted by Julius Caesar in 48 BCE set the year at 365.25 days with a leap year every four years. It required some intricate mathematics; however, a Greek astronomer, Sosigenes of

---

[3] See *The Ancient Tradition of Geometric Problems*, Wilbur Knorr, Dover Publications, 1993.

Alexandria, designed the calendar. The Julian calendar lasted for 1,500 years and was replaced by the Gregorian calendar in 1582, which we use today. It is corrected for the length of the year, which is 365.242190 days and is decreasing. If we had to identify any important Roman mathematical contributions, we would have to say it was their number system. Although it was not as advanced as the ancient Babylonian number system, the Roman system became universal due to the vastness of the Roman Empire. It remained in use for over a thousand years (Figure 2.11.)

Figure 2.11   Clock face using Roman numerals.

One can still see its use today on the cornerstones of buildings marking the year they were built, on the hulls of ships indicating the water level, on the faces of analog clocks marking the hours, identifying kings or popes such as King Edward VII, or Pope Leo XIV, and in various other applications. The great mathematician, engineer, and scientist Archimedes was killed by a Roman soldier in 212 BCE during the Second Punic War. It was perhaps the most significant impact of the Roman civilization on mathematics.

We owe much to the ancient Greeks for their logical framework and their deductive approach to mathematical proof, which has provided the foundation for much of the development of mathematics since. Many of their ideas and sought-after truths were ahead of their time, taking thousands of years for the world to fully recognize and accept their value and resulting achievements. This is the reason why we employ many of the Greek letters for our mathematical concepts.

Chapter 3

The Emergence of Our Number System

The world's decimal number system, which we take for granted, is a relative newcomer in the history of mathematics. It went through many stages of development over many years and reached its present form and widespread acceptance only about 500 years ago. Compare this to the 5000 years before the present system was accepted, when civilizations used other number systems. One of the reasons for its late development is that earlier civilizations mainly used mathematics to solve practical problems of everyday life, so a simpler construct, which required only basic arithmetic and geometry, sufficed. Around 800 CE, there was a more widespread effort to understand and develop mathematical concepts beyond the basics. However, it took until the 16th century for the decimal system to become accepted throughout most of the world, as Roman numerals had dominated Western civilization for more than 1000 years. What uniquely characterizes our decimal system is that it is a positional number system containing only 10 symbols. These are the only symbols needed to express any number, no matter how large or small. This may not seem impressive to anybody today who grew up learning and using our decimal system at a young age. However, when you become aware of the long history of number development, you will then be able to appreciate the power and simplicity of a positional number system like ours. Chapter 1 explores the advanced base 60 number system of the Babylonians, which was exceptional for an ancient number system because it also used position to construct its numbers. However, most civilizations that followed did not adopt a positional number system. The Babylonians used only two symbols, which required representing large numbers with many characters and could be confusing at times. It did have symbols for fractions, but did not have a zero, which was a major drawback. Subsequent civilizations did not adopt this number system for several reasons. One being that each number needed to have a different name until you reached 60. Therefore, one had to memorize 59 number names. After 60 there was some name repetition. Furthermore, a Babylonian multiplication table from 1 to 60 would contain 1830 entries! It was not easy to memorize that table! Our decimal system has a clear pattern for number names after twenty, that is, twenty-one, thirty-one, forty-one, etc. This makes it easier to learn the number names, and it contains at most 66 products to memorize. Figure 3.1 shows the 66 products, eliminating those that are equal because of the commutative property of multiplication.

1	2	3	4	5	6	7	8	9	10	11	12
2	4	6	8	10	12	14	16	18	20	22	24
3		9	12	15	18	21	24	27	30	33	36
4			16	20	24	28	32	36	40	44	48
5				25	30	35	40	45	50	55	60
6					36	42	48	54	60	66	72
7						49	56	63	70	77	84
8							64	72	80	88	96
9								81	90	99	108
10									100	110	120
11										121	132
12											144

Figure 3.1   66 common decimal products.

As a result, civilizations that came after the Babylonians adopted number systems with smaller number bases of ten or twenty symbols inspired by our fingers and toes. However, there is nothing special about our base 10 and, as mentioned in Chapter 1, a duodecimal number system based on 12 would have some advantages over one based on 10, because 12 has more divisors than 10. If we were polydactyl, that is, had six fingers on each hand, we would probably have developed such a system. Curiously, an estimated occurrence of 0.3–3.6 per 1000 babies are born with one polydactyl hand, and several types of Fauna, particularly cats, are polydactyl. (See Figure 3.2.)

The ancient Egyptian number system, though it was based on 10, had its drawbacks. It contained only unit fractions and the fraction $\frac{2}{3}$, and did not contain a zero. Its symbols were cumbersome and required some effort to produce, while multiplication and division were not easy to do.

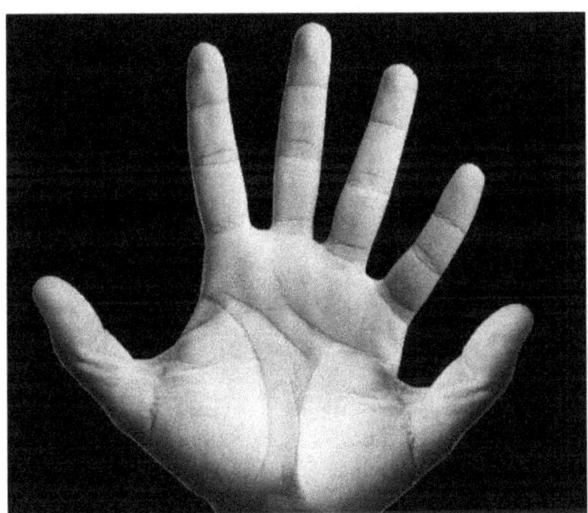

Figure 3.2   What if we had six fingers?

The Chinese characters used to represent numbers remained in use into the twentieth century, but only in China and some neighboring Far East countries such as Japan, Korea, and Vietnam. Not until the 1990s did China fully adopt the Hindu-Arabic numerals 0 to 9. The Maya Classical period lasted from about 250 BCE to 900 CE. It spread to more than 40 cities, each containing a population between 5000 and 50,000 people. Their number system, based on twenty, and quite advanced for the time, perished with their civilization. The ancient Greek number system, discussed in Chapter 2, and the Hebrew number system, which used letters to represent numbers, produced more than 27 symbols and were therefore not easy to work with. It satisfied the needs of money exchange, accounting, measuring, and other routine uses. However, Archimedes did create another number system capable of expressing vast numbers, but it was of little practical value.

The Roman numeral system was easier than the Greek system for doing basic arithmetic operations. It employed a limited number of symbols but was difficult to use for advanced mathematics beyond arithmetic. However, because of the large extent of the Roman Empire, it became universally adopted and trusted. As a result, it remained in use for over a thousand years after the end of the Roman Empire and was reluctantly replaced by our more advanced decimal system. Its reverence, though, has lasted up to the present day, and it is still used in many applications, such as cornerstones of buildings and other uses mentioned at the end of Chapter 2.

How then did our number system come to be? It is so familiar to us, and it is so ubiquitous, that it is hard to realize that its universal acceptance by the Western world is less than five hundred years old. The system bears the misleading name of Arabic. It is more properly called Hindu-Arabic, and there are still a few parts of the world that have not fully adopted it. Why our simple and easy-to-use number system took so long to achieve universal use should be of interest to all who use it every day and can then appreciate the effort it took to be adopted. It is such a versatile system, more than any other that existed before it, that no number is too great, or mathematical formulation too complex, to be expressed with its ten symbols. The development of the Hindu-Arabic system, like any great historical contribution, is subject to various origins. Here is a quote by the great French mathematician Pierre-Simon Laplace (1749–1827), who proved the stability of the solar system among his many contributions:

The ingenious method of expressing every possible number using a set of ten symbols (each symbol having a place value and an absolute value) emerged in India. The idea seems so simple nowadays that its significance and profound importance is no longer appreciated. Its simplicity lies in the way it facilitated calculation and placed arithmetic foremost amongst its useful inventions. The importance of this invention is more readily appreciated when one considers that it was beyond the two greatest men of antiquity, Archimedes and Apollonius.

We present here, as clearly as possible, from the evidence at hand, the story of how the Hindu-Arabic numerals came into being. We do know that the numerals 0, 1, 2, 3, 4, 5, 6, 7, 8, 9 closely resemble forms that were used in Europe in the fifteenth century. It was the printing press that helped spread the adoption of these symbols. It is interesting to note that many countries still use number symbols today, which are not the same as those presently used in Europe and the United States. Early evidence reveals that the Hindu-Arabic symbols may have evolved from number symbols used in the Brahmi writing system of South Asia as early as the third century BCE. Figure 3.3 shows Brahmi numerals from the first century CE. The number system was not a place-value system, and there were separate symbols for 10, 20, 30, 40, ..., 90, 100, 200, 300, ..., 1000, etc. Ancient Brahmi numerals have been discovered in caves and on coins in India, and it is found that they were in use for many years up to the fourth century CE. The Brahmi numerals for 1, 2, and 3 shown in Figure 3.3 are clearly understood, but it is not clear as to the origin of the other numerals. However, one can see some partial resemblance of these to our present-day numerals. One hypothesis is that the numerals may have come from the Brahmi alphabet. Some Brahmic scripts are still used today in parts of South and Southeast Asia.

1	2	3	4	5	6	7	8	9
—	=	≡	+	h	4	?	5	?

Figure 3.3   Brahmi numerals ca. 1st century CE.

The Brahmi numerals developed along several paths, one of which leads to our present system, which we shall explore. Around the fourth century CE, they first evolved into the Gupta symbols shown in Figure 3.4. The Gupta empire ruled in northeastern India from 400 CE to 600 CE, and its influence spread over a large area that the empire occupied. Beginning around 700 CE, the Gupta numerals were gradually replaced by the Nagari or Devanagari numerals, which spread from the civilization south of the Gupta empire (Figure 3.5).

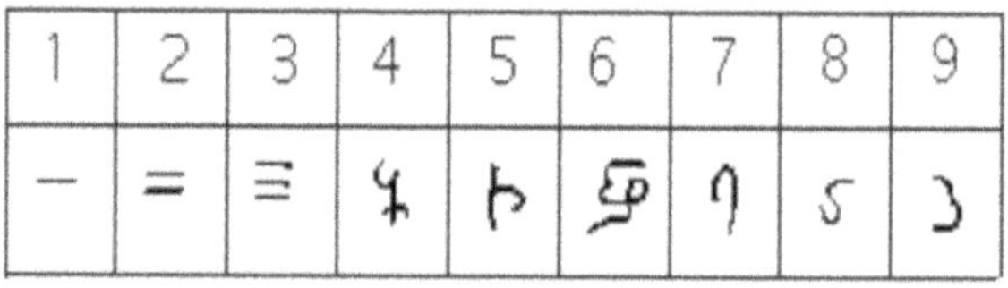
1
2
3
4
5
6
7
8
9

Figure 3.4 Gupta numerals ca. 4th century CE.

The name Nagari translates to "The writing of the Gods" as it was believed to be the most beautiful form to have evolved. The Iranian scholar and polymath Arab al-Biruni (973–1048), who has been called the "Father of modern geodesy," wrote:

> *What we use for numerals is a selection of the best and most regular figures in India.*

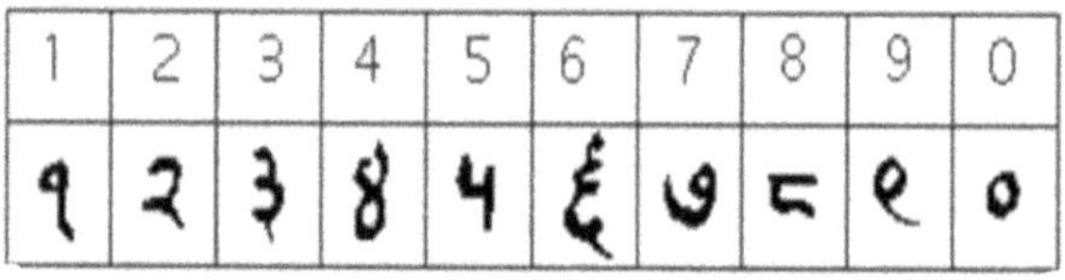
1
2
3
4
5
6
7
8
9
0

Figure 3.5 Nagari numerals ca. 11th century CE.

Note that the Nagari numerals for 1, 2, and 3 more closely resemble our numerals, and a zero has been added. The Indian number system was the most advanced system that existed throughout Asia and Europe at this time. It was a place-value system, that is, depending on the place of the numerals they took on different values. Though the Babylonians had a place-value system as early as 2000 BCE, it had a base of 60, while the Indian system was the first place-value system to have a base of ten. The Chinese "counting boards" discussed in Chapter 1 also had a base of ten, and evidence reveals that the Chinese system and the Babylonian system influenced the Indian system. It is remarkable, though, that it took almost three thousand years for another place-value system to emerge after the Babylonians. Besides developing the precursor of our present number system, a much more important contribution of the Indian civilization was the introduction of zero as a number and a place holder, which was almost non-existent in older number systems. It came into use as a number in Indian mathematics around 650 CE, and it is what makes the system so simple and yet so powerful. The idea of a number "zero" (The name comes from the Arabic *sifr*, which also means "cipher") is not an easily grasped concept as it conjures up the idea of "nothing", and therefore found little use in most ancient number systems. Older systems were used to solve concrete problems and lacked the level of abstraction that the concept of zero requires.[1] If ancient people wanted to communicate the number of goats or cows that they possessed, then the value 0 or −5 would not have a place in their number system. As sophisticated as the Babylonian number system was for its time, the absence of zero did not present a problem in their mathematics. The context in which the numbers were used helped to avoid confusion as to their value.

Before the symbol for zero came into use, Indian mathematicians would leave a space on their counting boards, and eventually this space became filled with subsequent calculations. The Indian term meaning void is *sunya*, which is how the Indians referred to zero. It is not known who invented the symbol for zero, but it seems to have evolved from a dot to a circle with a dot in it, and then to the oval shape we use today. An Indian document from 594 CE is the earliest evidence we have of a number written in the place-value form used today. Therefore, by the sixth century, the decimal place value system was fully in use in India. One of the speculations why the Indians may

---

[1] See "The History of Zero," pp. 9–12, *Mathematics: Its Historical Aspects, Wonders and Beyond*, Posamentier and Kramer, World Scientific, 2022.

have developed such a place-value system is their fascination with large numbers, which necessitated such a system. The founder of Buddhism, Gautama Buddha, was asked as a young man by his mentor to name all the numbers above a *koti*, which is 10^7. Buddha first listed all the powers of 10 up to 10^{53} and then proceeded to increase the values until he reached 10^{421}, which is an unbelievably large number, that is, 1 with 451 zeroes. His mentor then stated, *"You, not I, are the master mathematician."* This interest in large numbers resulted in naming successive powers of ten and may have led to a decimal place-value system where each place is a power of ten.

Another speculation is that the Indian system was influenced by the *Sand-reckoner* written by the Greek Archimedes, which included a system capable of expressing large numbers up to 8×10^{63}. However, as the Indian system developed, it gradually evolved and found its way into the rest of the world, forming the basis of the Western number systems. The transition began with the Arab countries in two directions. The western Arabic world, consisting of Spain and North Africa, embraced the Indian number system differently from the eastern Arabic world, and there was little exchange between the two groups. It was not a simple transition as it competed with several different mathematical systems employed by the Arabs at the time. One that was used by the academic community was a decimal place-value system that included fractions. Another system used primarily by the business community consisted entirely of words describing the numerals, and the fingers were used for counting. A third system used a sexagesimal base, where alphabetic letters were used to denote the numerals.

Before the Arab nations emerged, in 662 CE Bishop Severus Sebokht of the Byzantine Nestorian[2] church wrote about Indian computation being done with only nine symbols. This is one of the first signs that the Indian system had moved into the Western world. As the Arab empire grew, a reference to the Indian numerals being

---

[2] Nestorian Christians were followers of Nestorius of Constantinople, who believed that the human and the divine nature of Christ existed separately, as opposed to the Roman Catholic belief that they existed together in one person.

shown to a Caliph and disseminated to the Arabs is found in a famous work by the Egyptian Arab historian al-Qifti (1172–1248), *History of Learned Men.*

Al-Qifti writes that an Indian in 776 CE, who possessed a keen knowledge of astronomy, explained the movement of the heavenly bodies to the Caliph employing his advanced method of calculation. Astronomical calculations are mathematically demanding, and the Indian place-value system enabled these calculations to be performed more readily. The Caliph had the work translated into Arabic, as the location of the planets played an important part in the Arab calendar and religion. This astronomical work appears to be that of the Indian Brahmagupta (598–668), an accomplished mathematician and astronomer who was the first to clearly describe the quadratic formula to solve algebraic equations of the second degree. It was written in 628 CE using the Indian place-value system, making it evident that the Arabs had access to Indian numerals at this time. It is not clear exactly when the Arabs started writing and using the Indian numerals. Some claims are made that al-Khwarizmi (780–847), an Arab polymath, may have first employed them in one of his texts, but there remains a mystery here. Another Arabic book, *On the Use of the Hindu Numerals*, written around 830 CE by Al-Kindl (801–873), helped foster the adoption of the Indian symbols by the Arabs.

At first, the Indians did their calculations on a dust board, which, like a blackboard, enabled them to move numbers around and sometimes replace them with new results. However, this did not bode well for someone trying to follow the calculations, so written material soon replaced the dust board. The earliest surviving book that illustrates the Indian system was written around 952 CE by al-Uqlidisi (circa 920–980). In the book, he illustrates how practical the Indian system is:

> *Most arithmeticians are obliged to use it in their work: since it is easy and immediate, requires little memorization, provides quick answers and demands little thought.*

Al-Uqlidisi also promoted the use of pen and paper to replace the dust board, which enabled the adoption of the place-value system.

His work, besides helping to remove one of the obstacles to acceptance of the nine Indian number symbols, is also historically important as it was one of the first texts to work with decimal fractions. By the eleventh century, the Indian system became more widespread and was used by scholars, businessmen, and farmers. The number symbols were not the same in all regions and evolved. This was also the case in the Arabic world. Figure 3.6 shows the Indian numerals being used in the eastern Arabic empire around 969 CE.

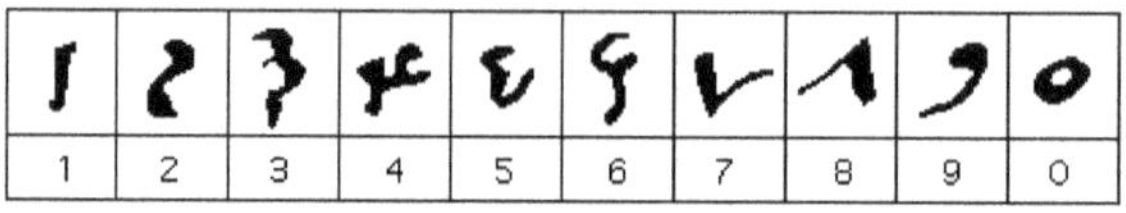

Figure 3.6   Eastern Arabic numerals ca. 10th century CE.

Compare these symbols to those used in the Western Arabic empire a few centuries later, which are shown in Figure 3.7. Not surprisingly, the Western symbols more closely resemble those that found their way into Europe.

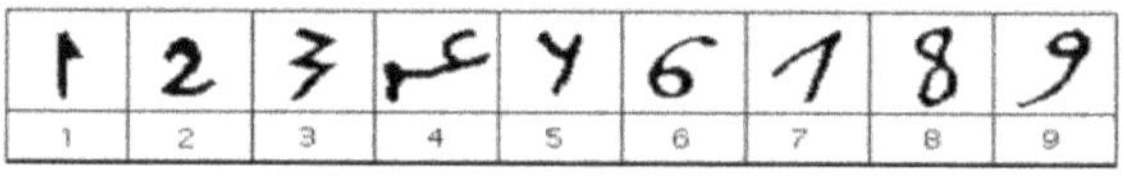

Figure 3.7   Western Arabic numerals ca. 14th century CE.

The earliest example of the Indian symbols appeared in the European document *Codex Vigilanus,* written in 976 CE by a monk in Spain. Europe, however, was still comfortable with Latin symbols and Roman numerals and was not ready to use new symbols or change its arithmetic calculations. Europe was slow to adopt the Hindu-Arabic system and continued using the Roman system as late as the 15th Century.

Enter Leonardo of Pisa (1170–1250), who was probably the most talented Western mathematician of the Middle Ages. He was the son of Guglielmo Bonacci and was also known as Leonardo Bigollo Pisano (Figure 3.8). Historians in the 19th century came up with the

Figure 3.8   Leonardo of Pisa (Fibonacci).

well-known name Fibonacci, which means "son of Bonacci.". Fibonacci was born in Italy, where his father, a Pisan merchant, was appointed consul for the merchants of the Republic of Pisa who were trading in the Mediterranean port of Bugia (now called Bejaia) in northern Algeria. Guglielmo, being a diplomat in North Africa, educated his son there. Fibonacci learned mathematics in Bugia from an Arab master and traveled extensively with his father. On his travels he was exposed to various mathematical systems and realized the many advantages of the Arabic numerals. Fibonacci writes in his famous book *Liber Abaci*, published in Pisa in 1202:

When my father, who had been appointed by his country as public notary in the customs at Bugia acting for the Pisan merchants going there, was in charge, he summoned me to him while I was still a child, and having an eye to usefulness and future convenience, desired me to stay there and receive instruction in the school of accounting. There, when I had been introduced to the art of the Indians' nine symbols through remarkable teaching, knowledge of the art very soon pleased me above all else and I came to understand it, and its use for what was studied by the art in Egypt, Syria, Greece, Sicily and Provence, in all its various forms.

Not much is known about Fibonacci's life except what is contained in his mathematical writings. When he was fifty years old, Fibonacci stopped traveling and settled in Pisa, where he wrote several important texts. His books made significant mathematical contributions and helped to increase the knowledge of ancient mathematical skills. During his time, there was no printing press, so only handwritten copies of his texts survived. The existing copies are *Practica Geometrie* (1220, "Practice of Geometry"), *Flos* (1225, "Flower", figuratively "the best"), *Liber Quadratorium* (1225, "Book of Square Numbers"), and his most famous, *Liber Abaci* (1202, "Book of Calculation"). Not a lot of scholarship flourished in Europe during Fibonacci's time but still his work attracted widespread interest. He was a very gifted mathematician and discovered many abstract concepts; however, it was his more concrete ideas that generated the most interest and contributed to his fame, as they had a significant impact on the arithmetic of the time.

In 1202, only a few learned men in Europe were familiar with the Hindu-Arabic numerals from translations of 9th-century Arab mathematicians, primarily al-Khwarizmi. The Arabic books explained the Hindu notation, the principle of place value involving units, tens, hundreds, etc., and how the numbers simplified arithmetic calculations. The solution of practical problems in the books, such as money changing, bartering, profit margins, interest, and the conversion of weights and measures, illustrated how much easier and effective Hindu-Arabic numerals were compared to more cumbersome medieval techniques.

It was Fibonacci who introduced the Hindu-Arabic numeral system in *Liber Abaci*. It was called *modus indorum* (method of the Indians), and it explained the positional notation with ten digits, including zero. The introduction of zero (*zephir* in Arabic) was a major accomplishment of the Indians, and it is what made the system so promising and successful. The book applies the number system to many of the same types of problems found in the Arab texts. The original manuscript of *Liber Abaci* was written in 1202 and unfortunately does not exist today. The second revised edition was produced in 1228, and also begins with the following:

The nine Indian[3] figures are: 9 8 7 6 5 4 3 2 1.[4]

With these nine figures, and with the sign 0, which the Arabs call zephyr, any number whatsoever is written, as demonstrated below. A number is a sum of units, and through the addition of them the number increases by steps without end. First, one composes those numbers, which are from one to ten. Second, from the tens are made those numbers, which are from ten up to one hundred. Third, from the hundreds are made those numbers, which are from one hundred up to one thousand. … and, thus, by an unending sequence of steps, any number whatsoever is constructed by joining the preceding numbers. The first place in the writing of the numbers is at the right. The second follows the first to the left.

The first section of *Liber Abaci* shows methods for converting between other number systems, including Roman numerals. It also explains how to test whether a number is a composite number, that is, whether it has divisors, and shows how to factor such a number into its divisors.

Figure 3.9 shows a page of Fibonacci's *Liber Abaci*. The box on the right shows the famous Fibonacci sequence, explained below. The position in the sequence is labeled with Latin numbers and Roman numerals, and the value in Hindu-Arabic numerals.

The second section in the 1228 version deals with monetary problems involving currency conversions and how to calculate profit and interest, which was very important to the banking industry. The third section discusses some mathematical problems involving prime

---

[3] Fibonacci used the term Indian figures for the Hindu numerals.

[4] It is assumed that Fibonacci wrote the numerals in order from right to left, since he took them from the Arabs who write in this direction.

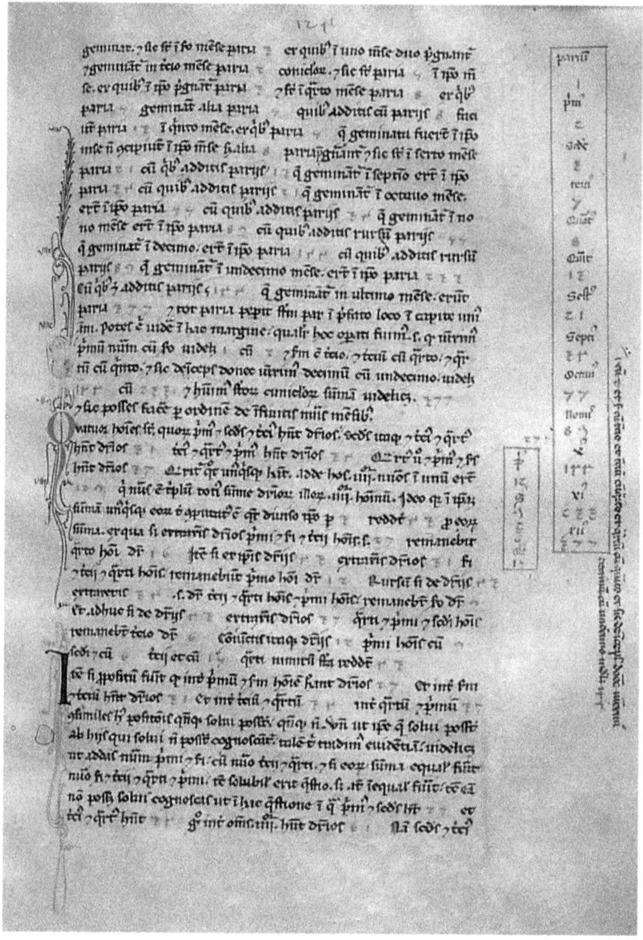

Figure 3.9   A page of Fibonacci's *Liber Abaci* from the Biblioteca Nazionale di Firenze.

and composite numbers and sequences of numbers. The famous Fibonacci sequence (Figure 3.10) is illustrated in Chapter 12. However, this sequence did not originate with Fibonacci. It was already described by Indian mathematicians around 600 CE, but was introduced to Western Europe as a result of his book and was, therefore, named after him. The Fibonacci sequence begins with two 1's, and then each successive number is obtained by adding the previous two numbers as follows:

1, 1, 2 (1+1)**, 3** (1+2)**, 5** (2+3)**, 8** (3+5)**, 13** (8+5) **... etc.**

Figure 3.10   Fibonacci sequence.

Fibonacci omitted zero, and the first 1, but the first 1 is included for completeness today. In *Liber Abaci* the sequence emanates from a problem in Chapter 12, which involves the procreation of rabbit couples. It states the following sequence:

For every pair of newly born rabbits in a field, they mate after one month and produce another pair of rabbits after two months. Then after every following month they continually produce another pair of rabbits. All the rabbits remain alive and never die.

Starting with one pair of rabbits, we have two pairs after two months. After three months, the first pair produces another pair, the second pair does not produce any, and so we have three pairs of rabbits. After four months, the first and the second pair each produce another pair, while the third pair does not produce any. So we now have 5 pairs. Continuing in this way, Fibonacci generates his sequence 1, 2, 3, 5, 8, 13, 21, 34, 55, 89, 144, ... etc., which is shown with the rabbits in Figure 3.11. Fibonacci posed the query: How many pairs will there be after one year? Can you answer that?[5] The name "Fibonacci sequence" was first used by the 19th-century French number theorist Édouard Lucas (1842–1891). *Liber Abaci* also includes irrational numbers in the fourth section, such as square roots and cube roots. They are shown as both numerical and geometrical approximations. There are also some proofs in Euclidean geometry and methods of solving algebraic equations.

---

[5] The answer is 233, which is the 12th term in the sequence.

Figure 3.11   Fibonacci's rabbits.

The Holy Roman Emperor Frederick II learned of Fibonacci's work from court scholars who had corresponded with Fibonacci. The court astronomer, Dominicus Hispanus, suggested to Frederick II that he invite Fibonacci to his court when it meets in Pisa around 1225. A member of Frederick II's court, Johannes of Palermo, an imperial notary, presented several problems as challenges to the great mathematician Fibonacci at his court appearance. Three of these problems were solved by Fibonacci and he presents solutions in his book *Flos*, which he sent to Frederick II. The first two were of a type developed by the third-century Greek mathematician Diophantus (ca. 200–284 CE). They are called Diophantine equations, which are equations that have several solutions that are whole numbers, and were a favorite of the Arabs.

The third problem Johannes challenged him with was to find the root of the cubic equation: $x^3 + 2x^2 + 10x = 20$. Fibonacci explains that the root is not an integer nor a fraction, but an infinite decimal, which is an irrational number. He gives an accurate approximation using a trial-and-error method in *Flos* to find one irrational root. The root was given in sexagesimal (base 60) notation as follows:

$$1 + \frac{22}{60} + \frac{7}{3,600} + \frac{42}{216,000} + \cdots \approx 1.368808108\ldots$$

The accuracy of the decimal equivalent is incredible. It is correct to nine decimal places, attesting to Fibonacci's genius! There are two other roots, which were not possible for Fibonacci to find, as they are *complex* numbers, meaning they contain the imaginary number $i = \sqrt{-1}$. Such numbers were not considered until the Renaissance period, several centuries later.

When educated Europeans read copies of *Liber Abaci*, they were profoundly impressed. The Hindu–Arabic numerals gradually replaced Roman numerals and other awkward arithmetic processes. It gradually replaced the use of the abacus. Business calculations were made easier and faster, resulting in the growth of banking and accounting throughout Europe.

After 1228, there is only one known document that refers to Fibonacci. This is a decree made by the Republic of Pisa in 1240, which honored Fibonacci (referred to as Leonardo Bigollo) by granting him a salary in recognition of the services that he had given to the city as an advisor on matters of accounting and instruction to citizens. In the 19th century, a statue of Fibonacci was erected in the western gallery of the Campo Santo, which is a historical cemetery located in the Cathedral Square in Pisa (Figure 3.12).

The Fibonacci sequence has been found to illustrate many natural phenomena such as the leaves on a stem, the flowering of an artichoke, the branches of a tree, the arrangement of a pinecone's bracts, and the spirals on a nautilus seashell, to name a few. (See Figure 3.13.) In 1982, an asteroid belt was discovered and named *6765 Fibonacci*, as this number is the sum of the 18th and 19th Fibonacci numbers, that is, $4181 + 2584 = 6765$.

There are many applications in mathematics for the Fibonacci numbers.[6] One that is particularly interesting is the Golden Ratio. This is a rectangle whose shape is considered most appealing and has the following ratio of its sides:

$$\frac{Length}{Width} = \frac{Length + Width}{Length}$$

---

[6] See "The Fibonacci Numbers," pp. 333–340, *Mathematics: Its Historical Aspects, Wonders and Beyond,* Posamentier and Kramer, World Scientific, 2022. Also see *The Fabulous Fibonacci Numbers*, Posamentier and Lehmann, Prometheus Books, 2007.

Figure 3.12   Fibonacci at the gallery of the Campo Santo.

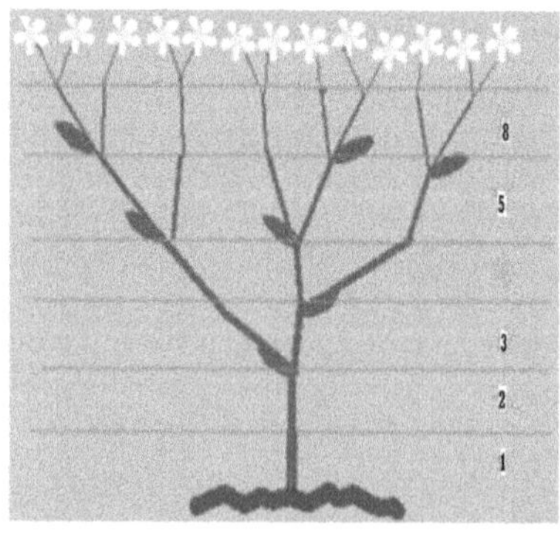

Figure 3.13   Fibonacci sequence in a tree and nautilus shells.

The golden ratio can be readily calculated by assuming $w = 1$ and writing the equation:

$$\frac{l}{1} = \frac{l + 1}{l}$$

This simplifies to the quadratic equation: $l^2 = l + 1$

The quadratic formula then yields the solution: $l = \frac{1+\sqrt{5}}{2} \approx \mathbf{1.618.}$

That is, the length of the golden rectangle is approximately 1.6 times its width.

This shape has been appreciated for centuries and can be found in paintings and the structure of buildings such as the Parthenon (Figure 3.14) in Greece. The ratio of two very large consecutive Fibonacci numbers approximates the Golden Ratio. For example, $\frac{1597}{987} \approx 1.618$, where 987 and 1597 are numbers in the Fibonacci sequence. The larger the two consecutive Fibonacci numbers are, the closer their ratio approximates the Golden Ratio. Leonardo of Pisa forever changed Western methods of calculation and created challenges for mathematicians to this day. The Fibonacci Association was started in 1963 as a tribute to the master mathematician and publishes the *Fibonacci Quarterly* to continuously enhance our knowledge of the Fibonacci numbers.

As an epilogue to this story, we move to 1790, the time of Napoleon, more than five centuries after Fibonacci introduced the base-ten system to Europe. At this time, we still find the vestiges of ancient number systems that were not based on ten being used in the measurement systems of England, the United States, and many other countries. Finally, thanks to Napoleon, these old number systems were replaced in France by the decimal metric system. The Système International (SI system), as the metric system is called today, took another two centuries to be almost universally adopted. Only three countries, the United States, Myanmar, and Liberia, still use the US customary system, which had its origins in the old British system. Most industries in the United States use the SI system as they need to conform to international standards. On 23 December 1975, President Gerald R. Ford signed the *Metric Conversion Act of 1975*, which attempted to give official federal sanction for the US to convert to the

Figure 3.14   Golden rectangle in the Parthenon.

metric system. This 1975 law established the 17-member United States Metric Board (USMB) to coordinate the voluntary conversion to the metric system. Unfortunately, people were so used to our US customary system that they were not comfortable with metric measures. So much written material needed to be revised that the Board was dissolved in 1982. So many years later, everyday consumers still cling to the old measures.

We now realize that our number system, which we take for granted and find so easy to use, took many centuries to evolve, even though it was introduced to the Western world by Fibonacci's book over a thousand years ago. This should engender a great appreciation for the remarkable achievement it is, which is a testimony to its simplicity and beauty. Mathematicians consider the greatness of a mathematical discovery to lie in its simplicity and beauty. Here are two examples of such exquisite mathematical beauty:

Einstein's famous concise equation relating mass m and the speed of light c to energy:

$$E = mc^2$$

And Euler's most remarkable equation:

$$e^{\pi i} + 1 = 0$$

Euler's equation relates five basic mathematical constants: Euler's number $e \approx 2.71828$,

$$\pi \approx 3.14159, \text{ the imaginary unit} = i, 0 \text{ and } 1.$$

It has been considered the most beautiful equation in mathematics. Unfortunately, the meaning of this equation cannot be fully discussed here, but it is shown and hopefully appreciated for its mathematical elegance.

Chapter 4

The Origin and Growth of Algebra

The Medieval Era saw the rise of the Islamic armies of Muhammad, who occupied Mecca and, within a century, much of the Middle East, Northern Africa, and parts of India and Spain. Their learned mathematicians gave momentum to the place-value decimal system and provided the mathematical foundation that gave birth to algebra. Greek and Indian works were translated into Arabic, and by the ninth century, the mathematics of Euclid, Archimedes, Apollonius, Ptolemy, and other Greek mathematicians were all accessible to the Arabs. Many Muslims believed that discovering mathematical ideas that were beyond utilitarian value was one of the dictates of their God, Muhammad.

We owe the word algebra and its beginnings to the Islamic mathematician al-Khwārizmī (780–850 CE) (Figure 4.1). His treatise *Al-Kitab al-Mukhtasar fi Hisab al-Jabr al-Muqābala* ("The Compendious Book on the Calculation of al-Jabr and al-Muqābala") was the seminal book written on algebra in 825 CE. The terms *al-jabr* and *al-muqābala* mean, respectively, "restoring" and "comparing" and refer to the operations of addition and subtraction that can be done equally on both sides of an equation. Changing the equation $x - 2 = 5$ to $x = 7$ is an example of *al-jabr*, where 2 is added to each side of the equation, whereas the conversion of $x + 2 = 5$ to $x = 3$ is an example of *al-muqābala*, where 2 is subtracted from both sides of the equation. The word "algebra" developed as an adaptation of the Arabic *al-jabr*. What makes al-Khwārizmī's book a most significant contribution was that it constituted a major move away from geometry, which was the direction of mathematics fostered by the Greek civilization. Algebra allowed the unification of almost all mathematical ideas. Numbers, arithmetic processes, Euclidean geometry, and trigonometry could all be treated with algebraic expressions and equations. It provided for the expansion of mathematical concepts and paved the way for the continual development of ideas. As the Egyptian mathematician Roshdi Rashed writes in his book[1]:

> *Al-Khwārizmī's successors undertook a systematic application of arithmetic to algebra, algebra to arithmetic, both to trigonometry, algebra to the Euclidean theory of numbers, algebra to geometry, and geometry to algebra. This was how the creation of polynomial algebra, combinatorial analysis, numerical analysis, the numerical solution of equations, the new elementary theory of numbers, and the geometric construction of equations arose.*

Figure 4.1   Monument of al-Khwarizmi in Khiva, Uzbekistan.

---

[1] *The Development of Arabic Mathematics Between Arithmetic and Algebra*, Rashed, Springer Nature, London 1994.

Al-Khwārizmī's work, though somewhat theoretical, also proved to be very practical in solving problems in business, land measurement, and architecture. He created his own terminology for algebraic expressions, distinguishing them from arithmetic terms, and demonstrated algorithms and mathematical operations using only words and not symbols. He first applied his ideas to linear and quadratic equations, and to operations on binomials and trinomials. An amazing achievement was his solution for the general quadratic equation in the form: $ax^2 + bx = c$. Today, this equation would be written $ax^2 + bx - c = 0$, but negative quantities were not considered acceptable numbers by al-Khwārizmī, which stemmed from his following of Greek mathematics. He did accept negative quantities in certain contexts because of the work of the Indian mathematician Brahmagupta (598–670 CE), who applied them to financial dealings in 620 CE. Brahmagupta devised the following rules for operating with positive and negative numbers, which he called fortunes and debts, respectively, and the number zero, which the Indians were the first to use as a placeholder and a number:

> *A debt minus zero is a debt.*
> (Negative minus zero is negative.)
> *A fortune minus zero is a fortune.*
> (Positive minus zero is positive.)
> *Zero minus zero is a zero.*
> *A debt subtracted from zero is a fortune.*
> (Zero minus negative is positive.)
> *A fortune subtracted from zero is a debt.*
> (Zero minus positive is negative.)
> *The product of zero multiplied by a debt or fortune is zero.*
> (Zero times negative or positive is zero.)
> *The product of zero multiplied by zero is zero.*
> *The product or quotient of two fortunes is one fortune.*
> (Product or quotient of positive numbers is positive.)
> *The product or quotient of two debts is one fortune.*
> (Product or quotient of negative numbers is positive.)
> *The product or quotient of a debt and a fortune is a debt.*
> (Product or quotient of a negative number and a positive number is negative.)

The product or quotient of a fortune and a debt is a debt.
(Product or quotient of a positive number and a negative number is negative.)
The quotient of zero and zero is zero.
(Zero divided by zero is zero.)

It is truly impressive that these rules, formulated more than 1700 years ago, are exactly what we use in algebra today, except for the last one, which is not correct. Understandably, Brahmagupta might have come up with this rule as he wanted to include all the possible arithmetic combinations involving zero. However, zero divided by zero is not equal to any specific number; that is, it is undefined. You are always told in school that you cannot divide by zero, but you may not be shown why. The reason is that we have a rule that any number N times zero is equal to zero: $N \times 0 = 0$. If we want this rule to be true, then it must follow that $\frac{0}{0} = N$, because division is the inverse of multiplication. That is, if $A \times B = C$, then $\frac{C}{B} = A$. Now, if $\frac{0}{0} = N$, and N can be any number, then zero divided by zero can be any number N. Therefore, we must accept that zero divided by zero is undefined, and we cannot permit division by zero for any operation. If we did permit it, it would be possible to prove that any number is equal to any other number. Consider any two numbers a and b. It is true that: $a \times 0 = b \times 0$.

Then, if one divides both sides by zero, we have:

$$\frac{a \times 0}{0} = \frac{b \times 0}{0}$$

Now if we are permitted to divide out the zeros we would have:

$a = b$, which is clearly incorrect

Now let us return to al-Khwārizmī's treatment of the quadratic equation, which he necessarily divided into five types because he only considered positive quantities for all the terms and the solutions:

$$ax^2 = bx$$
$$ax^2 = c$$
$$ax^2 + bx = c$$
$$bx + c = ax^2$$
$$ax^2 + c = bx$$

The solutions to the first two equations are somewhat straightforward. Zero is not considered a solution to the first type, so you can divide by x on each side and obtain: $ax = b$. Then the one root is $x = \frac{b}{a}$. The solution to the second type also has only one root $x = \sqrt{\frac{c}{a}}$ because Arab mathematicians did not accept a negative solution. It is, however, interesting to note that the Hindus did accept negative solutions in their mathematics. The last three equations above are present in today's algebra because we accept negative terms. So, let's look at one of them, for example, the third type, $ax^2 + bx = c$. It would have been done by al-Khwārizmī using only words, since he did not use symbols. The equation $x^2 + 6x = 16$ would appear as: "What value when squared and increased by six of its roots sums to 16?" His solution is as follows: Take half of the number of roots ($\frac{6}{2} = 3$) and square this value ($3^2 = 9$). Add this to the number on the other side of the equation ($9 + 16 = 25$) and then take its square root ($\sqrt{25} = 5$). Now obtain the solution by subtracting half the number of roots from this final value ($5 - 3 = 2$). The one positive solution is then 2.

The method that al-Khwārizmī describes is the same one we teach today as a part of elementary algebra and is called "completing the square." Here is how it would work for this example. Starting with the equation $x^2 + 6x = 16$, we take half of 6 and square it to get 9. We then add it to both sides of the equation: $x^2 + 6x + 9 = 16 + 9 = 25$. Now, before we take the square root of 25, we factor the left side of the equation to obtain: $(x + 3)^2 = 25$. Then we can take the square root of both sides of the equation as follows: $\sqrt{(x+3)^2} = \sqrt{25}$. This simplifies to $x + 3 = 5$. We now do the last step described in the solution above and subtract 3 from both sides to obtain the positive solution $x = 2$.

Al-Khwārizmī's solution of quadratic equations was a major mathematical achievement and set the groundwork for the further development of not just quadratic equations, but also polynomial equations involving higher powers of x. Among many other types of problems in his text, he clearly stated the rule for multiplying binomials, considering both positive and negative values. For example, he stated that if a binomial $(a + b)$ is multiplied by another binomial

$(c + d)$. then four multiplications are necessary: $ac + ad + bc + bd$. Furthermore, if b and d are positive, then bd is positive. If one of them is negative, then bd is negative, and if both are negative, then bd is positive. In a few rare examples, he considered solutions that are square roots and not rational, that is, not fractions or whole numbers. One such example states, "If I divide 10 into two parts, such that ten times one part equals the square of the other part, what are the two parts?" If p is one part, then $(10 - p)$ is the other part. Then $(10)(10 - p) = p^2$. This leads to the quadratic equation: $p^2 + 10p = 100$, and the positive solution is $p = -5 + 5\sqrt{5}$ which is approximately 6.18 and $10 - p \approx 3.82$.

To check the solution: $(6.18)^2 = 38.2$ and $10 \times 3.82 = 38.2$.

Many of the problems in his book deal with quadratic equations, but only some were of practical value, such as this example: "One *dirhem* (a unit of Arabic currency) is divided among several men. Suppose then one man joins the group, and you divide one dirhem again among the increased group. If each man receives one-sixth of a dirhem less the second time, how many men were in the original group?" This problem requires a thoughtful application of algebra. If n = the number of men in the original group, then at first each man receives $\frac{1}{n}$ of a dirhem. The second time, each man receives $\frac{1}{(n+1)}$ of a dirhem. The problem states that the last amount is $\frac{1}{6}$ of a dirhem less than the first amount, so if we subtract the amounts, the equation is: $\frac{1}{n} - \frac{1}{(n+1)} = \frac{1}{6}$. This simplifies to $n^2 + n = 6$, and the solution is $n = 2$.

Understandably, al-Khwārizmī could not include many practical applications of quadratic equations in his book because even today, when you consider the everyday needs of most people, few problems in their lives require the need to solve quadratic equations. Nevertheless, after his book was translated into Latin during the 12th century, it played a pivotal role in the dissemination of algebraic knowledge throughout Europe. This had a significant impact on mathematicians during the Renaissance and helped shape the evolution of modern mathematics. The contributions of al-Khwārizmī exemplify the rich mathematical knowledge of the Islamic world and its enduring impact on Western mathematics.

Figure 4.2   Abu Kamil Shuja.

One of the brightest mathematicians who perfected al-Khwārizmī's work on algebra was the Egyptian mathematician Abu Kamil Shuja (850–930 CE; Figure 4.2), who produced solutions to more complex problems by dealing more effectively with irrational quantities. Abu Shuja was born in Egypt the same year al-Khwārizmī died and is considered his successor, by publishing treatises in arithmetic, algebra, logic, astronomy and land surveying. His texts were some of the best-selling books of his era, and they became one of Fibonacci's main sources in transmitting the Arab work throughout Medieval Europe. Abu Kamil deserves much of the credit for bringing the decimal place-value system and the fundamental ideas of algebra to the Western world.

After Abu Kamil, much work by talented Arabic mathematicians continued for more than two centuries, applying the basics of algebra to many of the geometric problems of the Greeks. They developed the algebra of operations with polynomials, solved certain cubic equations, calculated the volumes of solid figures applying the Greek mathematician Eudoxus's method of exhaustion (a forerunner to today's calculus), and derived the binomial coefficients for the powers of a binomial. The array of binomial coefficients shown in the boxes in Figure 4.3 is called Pascal's triangle. They are shown for the first four binomials from $(a + b)^1$ to $(a + b)^4$. It is called Pascal's triangle because the French mathematician Blaise Pascal (1623–1662) showed how these coefficients can be arranged in a triangle and are fundamental to the laws of probability. The first coefficients on either end are 1's, and each line of the remaining coefficients can be calculated from the previous line by adding the two numbers directly above it. For example, the 6 in $(a + b)^4$ can be obtained by adding the two 3's above it.

$$(a + b)^1 = \boxed{1}a + \boxed{1}b$$

$$(a + b)^2 = \boxed{1}a^2 + \boxed{2}ab + \boxed{1}b^2$$

$$(a + b)^3 = \boxed{1}a^3 + \boxed{3}a^2b + \boxed{3}ab^2 + \boxed{1}b^3$$

$$(a + b)^4 = \boxed{1}a^4 + \boxed{4}a^3b + \boxed{6}a^2b^2 + \boxed{4}ab^3 + \boxed{1}b^4$$

Figure 4.3   Pascal's triangle.

To obtain the coefficients for $(a + b)^5$, start with the line of coefficients for $(a + b)^4$: 1, 4, 6, 4, 1, The next line would then be 1 5 10 10 5 1. Applying this reasoning what would be the coefficients for $(a + b)^{6}$?[2] Pascal's contributions to mathematics are discussed further in Chapter 5.

Some of the very important contributions by Islamic mathematicians were advances in spherical trigonometry. This material was necessary for Muslims because they needed to know the *qibla*, the direction of Mecca that a Muslim must face during prayer. The method to calculate the *qibla* was discovered during the tenth century, and by the thirteenth century, methods for solving all types of plane and spherical triangles had been developed. They were published by the Muslim Nasīr al-Dīn al-Tūsī (1201–1274; Figure 4.4), a very wise Persian polymath. Some consider him the creator of trigonometry from his comprehensive work *Treatise on the Transversal Figure*. He

---

[2] 1, 6, 15, 20, 15, 6, 1.

published many works in mathematics, engineering, astronomy, and logic and is considered one of the most outstanding scientists of medieval Islam.

Figure 4.4   Nasīr al-Dīn al-Tūsī.

Meanwhile, in Spain, Euclid's *Elements* became available in Latin early in the twelfth century, and Spanish and French Jews made significant contributions in mathematics, drawing mainly from the work of Islamic scholars. Abraham bar Hiyya (1070–1136) of Barcelona published a work in Hebrew in 1116 to help Spanish and French Jews measure their agricultural fields. In his *Treatise on Mensuration and Calculation*, he provided rules for finding areas and volumes of various geometric figures. One of his original contributions has to do with the circumference, C, and the area, A, of a circle. He gives the standard formulas: $C = 2\pi r$ and $A = \pi r^2$, where r is the radius of the circle. He notes that the value of π that can be used is $3\frac{1}{7}$, which is the value used by most students when they first encounter the circle in basic geometry. He then states that a more exact value one can use in astronomical calculations, which requires more precision, is $3\frac{17}{120} = 3.1416667$. This is quite remarkable as his value is closer to the actual value of $\pi = 3.1415926$ than the range of values determined by Archimedes, who found that π was between $3\frac{10}{71} = 3.1408451$ and $3\frac{1}{7} = 3.1428571$. Abraham bar Hiyya also gives the formula for the area of a circle in the form: $A = \frac{Cr}{2}$. This formula is correct if you substitute $2\pi r$ for C so that you obtain $A = \frac{(2\pi r)r}{2} = \pi r^2$. He justifies his formula by first thinking of a circle that is flexible, then slicing it along the radius, opening it up, and flattening it out into a triangle, as shown in Figure 4.5. This, of course, is physically impossible, but if one imagines it could be done, then the area of the triangle would equal the area of the circle. The area of the triangle is one-half the base times the height: $A = \frac{1}{2}Cr = \frac{1}{2}(2\pi r)(r) = \pi r^2$, which is the correct formula for the area of a circle!

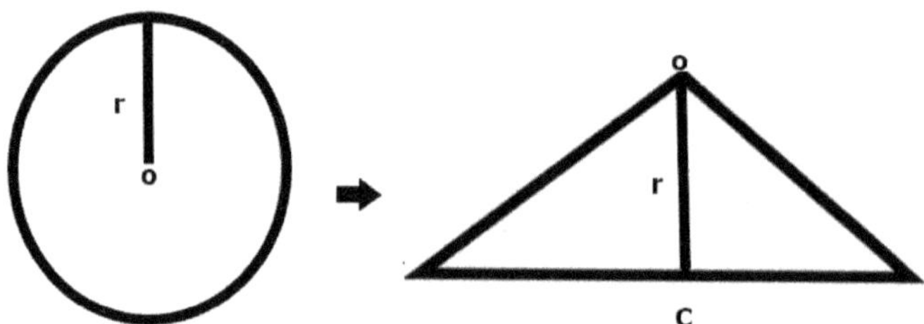

Figure 4.5   Circle opened up into a triangle.

Another prominent Jewish mathematician, philosopher, theologian, and astronomer of the Middle Ages was Levi Ben Gerson (1288–1344). He was born in Bagnols-sur-Cèze, France, and is better known by the Greek form of his name: Gersonides. His theorems on combinatorics, ways to arrange and combine different elements, were proven quite rigorously and represent a significant mathematical contribution in addition to his other important ideas in the fields of religion and philosophy.

Eventually, the Golden Age of the Islamic Empire diminished as the Europeans defeated its armies and regained control of its lands. The scholars in Europe mastered the work of the Arabs, who became less motivated to do scientific inquiry and tightened their religious bonds. However, progress was slowed down by the Hundred Years' War (1337–1453) and the Black Death (1446–1453), which killed a third of the European population. The Spanish Inquisition in 1492 also dealt a blow to the Jews in Spain and France.

The Renaissance

As the Medieval Era gradually faded, Europe began to wake from its long sleep and perceive faint glimmers of the coming Renaissance. The printing press was invented, resulting in the widespread dissemination of knowledge, and the Church became less of a dominant force. The work of Fibonacci, instrumental in the adoption of the place value decimal system in Europe, and the creation of the foundations of algebra, enabled mathematics to flourish, along with the advancement of science and technology. Nowhere else during the fifteenth and sixteenth centuries did the heart of the European Renaissance blossom as much as in the Italian states. They provided support to talented individuals with the wealth gained through their network of merchants and their many trade routes to the East. Credit is due to Marco Polo (1254–1324), the Venetian merchant, explorer, and writer, who ventured along the great Silk Road to China and the far East. Mathematics, science, and art thrived in Italy during the Renaissance, and Giotto di Bondone (Giotto) changed the course of painting and freed it from Byzantine conventions, spawning the genius of Leonardo da Vinci and Michelangelo.

As the economy became more modernized, the writing of checks was introduced, replacing the old method of bartering, and a class of professional mathematicians called "abacists" appeared. The abacists taught the merchants the place-value decimal system and how to apply it. However, there remained resistance to the new methods, and the use of Roman numerals remained for many years. Eventually, with the teaching of the new methods to children, the decimal system was adopted throughout Europe. Though much of the teachings of the abacists were for practical applications, they did introduce symbols and abbreviations which the Islamic algebra did not employ, adding enrichment to the subject and allowing for further inquiry and advancement.

One of the most comprehensive mathematics texts printed during this time was written by Luca Pacioli (1445–1517) and titled *Summa de Arithmetica, Geometria, Proportioni et Proportionalita*.[3] He was a

---

[3] First printed in Venice, 1494.

Franciscan friar and a mathematician who lived with Leonardo da Vinci while giving him mathematics lessons. Though his mathematics text was mostly a compendium of previously established material, the fact that it was printed contributed to it being widely circulated and well-studied by mathematicians during the sixteenth century. As a result, it helped contribute to many of the advances in algebra during the Renaissance.

Luca Pacioli also published a work on accounting and the double-entry method of bookkeeping, earning him the title "Father of Accounting". Another one of his books on mathematical and artistic proportion, *Divina Proportione*, introduces the golden ratio and includes illustrations by Leonardo da Vinci of the regular solids. One is shown in Figure 4.6. The use of perspective, which began to be used by artists such as Piero della Francesca and Marco Palmezzano at the time, is also discussed in the book.

During the Renaissance, the positive solutions to all the forms of the general quadratic equation $ax^2 + bx = c$ were well known, and many mathematicians were then challenged to find the positive algebraic solutions to all the forms of the general cubic equation: $ax^3 + bx^2 + cx = d$. Because most mathematicians still did not deal with negative numbers, there were thirteen forms of the cubic equation that had to be considered, depending on the relative positions of the positive terms.

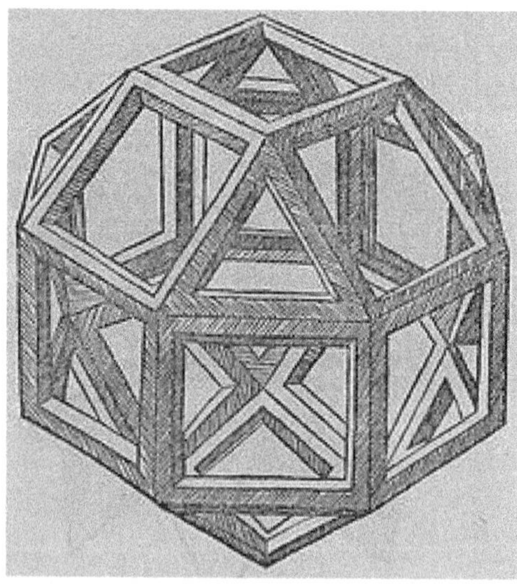

Figure 4.6   A rhombicuboctahedron (18 squares and 8 triangles).

For example, one form, $x^3 + cx = d$, was first solved algebraically by Scipione del Ferro (1465–1526), a mathematics professor at the University of Bologna. This was one of the major ancient mathematical problems, and he deserves praise for his work, but unfortunately, he is not well known. However, del Ferro did not publish his solution since it was necessary to keep such a skill secret to help secure a university position. It was the procedure at the time for each contender for a position to challenge the other with problems to be solved, and then for each to present their solutions in a public forum. If a professor knew of a new method to solve certain problems, it was to his advantage to challenge his contender with such problems. Del Ferro was succeeded by Niccolò Fontana, known as Tartaglia (1499–1557), a talented mathematician born in Brescia who narrowly escaped several tragedies. When he was 6 years old, his father was murdered, and then when he was 12 years old, the French invaded his town, and he was almost killed by a sword that cut his jaw and palate. A beard always hid his unsightly scars, and he spoke with difficulty, earning the nickname *Tartaglia*, which means stammerer. Tartaglia had great ability and taught himself mathematics, but struggled for years as a teacher. In a public contest with a less capable mathematician, he successfully prevailed, having previously discovered the solution to the cubic of the form $x^3 + bx^2 = d$. What followed was a torturous encounter with the most influential mathematician of the day, Gerolamo Cardano (1501–1576) from Milan, Italy, who heard of Tartaglia's success. Cardano tried to extricate the solution from Tartaglia, who at first refused, but eventually conceded when Cardano promised to introduce him to the governor of Milan, which could help him obtain a better job. He revealed his solution with the pledge that Cardano would not publish it. Meanwhile, Cardano and his student Lodovico Ferrari (1522–1565) solved the thirteen cases of the cubic equation. They even found the solution to the twenty different types of the quartic, or fourth degree, equation. After Scipione del Ferro's death, Cardano inspected his papers and discovered that del Ferro was the first to solve a cubic equation. Feeling emboldened that Tartaglia was not the first to find a cubic solution, Cardano, in 1545, published his very significant work, *The Great Art of Algebraic Rules* (*Artis Magnae Sive de Regulis Algebraicis Liber Unus*, or simply *Ars Magna*) (Figure 4.7).

Figure 4.7 *Ars Magna* by Gerolamo Cardano.

The book was devoted to the solutions of the cubic and the quartic equations, with credit given to del Ferro, Tartaglia, and Ferrari for their discoveries. It was the first Latin treatise on algebra. Tartaglia was incensed that Cardano revealed his result and published a book, *New Problems and Inventions*, in which he told his side of the story. He insulted Cardano, who was now considered one of the world's leading mathematicians, and therefore, was able to disregard Tartaglia's invective. Cardano's associate Ferrari was angered by Tartaglia's comments and challenged him to a public debate in which Tartaglia tried to involve Cardano without success. Tartaglia was still struggling financially as a mathematics teacher in Venice. Ferrari, being very familiar with all the solutions in *Ars Magna*, managed to prevail over Tartaglia, who left before finishing the debate. As a result, Tartaglia lost an opportunity for a lectureship in his hometown of Brescia and returned to his job in Venice. He is credited with his discovery, as the formula for the cubic solution has been named the Cardano-Tartaglia formula. Though Tartaglia died a poor man, he is known for producing the first Italian translation of Euclid's *Elements* in 1543 and Latin editions of Archimedes' works.

Gerolamo Cardano, Tartaglia's rival, was a man of many accomplishments whose life was anything but dull. He was born out of wedlock, and his mother supposedly tried to abort him. His three older siblings died of the plague, and he was a sickly but very bright young man who earned a degree from the University of Padua. His father was a lawyer and a competent mathematics professor who died while Cardano was at the University and left him a small sum of money. After graduation, he survived as a gambler, being able to make a living by outwitting his opponents with his understanding of the foundations of probability. He was the first in the Western world to introduce the binomial coefficients and the binomial theorem, which are important in the theory of probability. Eventually, he earned a doctorate in medicine in 1525. Though a brilliant student, he was disliked for his arrogance and was refused admittance to the College of Physicians in Milan.

He tried setting up a medical practice but was unable to support his family. He then managed to obtain a position his father had held as a lecturer in mathematics at the Piatti Foundation in Milan. While a lecturer, he continued to medically treat people and achieved miracle cures, which led to him being highly regarded as a doctor and resulted in grateful support from his patients. Eventually, the College of Physicians recognized his ability and accepted him in 1539, the year he began a prolific publishing career writing on many topics, including medicine, philosophy, astronomy, and, of course, mathematics. Cardano published over 200 books on mathematics and science. Curiously, in 1540, he gave up his post at the Piatti foundation and devoted two years to gambling and playing chess most of the day. During the years 1543–1552, Cardano lectured on medical topics at the Universities of Milan and Pavia and also pursued mathematics publishing his great work *Ars Magna*. He found that the solutions to cubic and quartic equations produce at times roots of negative numbers. Such roots are complex numbers that were not clearly understood and not considered mathematical quantities at the time. In *Ars Magna* Cardano comes up with the complex roots $\left(5+\sqrt{-15}\right)$ and $\left(5-\sqrt{-15}\right)$ for a cubic equation. He multiplies them out to obtain the four products as follows:

$$\left(5+\sqrt{-15}\right)\left(5-\sqrt{-15}\right)=(5)(5)+\left(5\sqrt{-15}\right)+\left(-5\sqrt{-15}\right)\left(\sqrt{-15}\right)\left(\sqrt{-15}\right)$$

Now he cancelled the two inner products since one is positive and one is negative. The last product represents the square root of a number times itself. He knew that the square root of any positive number, when multiplied by itself, equals the number. For example: $\left(\sqrt{16}\right)\left(\sqrt{16}\right)=(4)(4)=16$ and in general $\left(\sqrt{X}\right)\left(\sqrt{X}\right)=X$. He applied the same reasoning to the last term for the negative number, and it resulted in just the number −15. The final product then becomes $25-(-15)=40$. This is exactly correct in today's use of complex numbers. Note that the notation used here is modern, and Cardano would have done this using less convenient notation involving words and letters. However, witnessing this curious answer, he dismisses his result and states:

> *and thus far does arithmetical subtlety go, of which this, the extreme, is, as I have said, so subtle that it is useless.*

His talent in medicine earned him the title of rector of the College of Physicians in Rome and he was considered one of the greatest doctors in the world and sought after by many heads of state. He cured the Archbishop of St. Andrews, who had suffered from asthma for ten years, and thereby achieved fame and fortune, but it was not to last. His eldest son, Giambattista, whom he held in high regard, married a shameless woman who extorted his money and bore children out of wedlock. Giambattista was enraged and fatally poisoned his wife. He was arrested and Cardano tried but could not prevent his execution. Cardano's professional life suffered from this experience, and his titles were lost. He was then faced with his youngest son Aldo gambling away all his possessions and much of his father's money. He never forgave his son Aldo. The church then imprisoned Cardano because one of his books contained heretical material about Jesus Christ. However, he was pardoned after only a few months and forgiven by the Pope. As compensation, he was awarded a pension and granted membership again in the College of Physicians in Rome. In his autobiography, he writes of his four greatest sadnesses:

The first was my marriage; the second, the bitter death of my son; the third, imprisonment; the fourth, the base character of my youngest son.

Gerolamo Cardano remains one of the great Renaissance masters of mathematics and science, whose *Ars Magna* substantially advanced the algebra of Islam, and it prevailed in Europe for many centuries. Fifteen years after the publication of *Ars Magna*, Rafael Bombelli (1526–1572; Figure 4.8), an Italian hydraulic engineer and a talented mathematician, felt that Cardano's exposition was most important but not well written, and not easily accessible for people who did not possess a good grasp of mathematics. He therefore proceeded to write a text in Italian called *Algebra*, clearly explaining many of Cardano's ideas so that most students could learn the material by themselves.

Figure 4.8   Rafael Bombelli.

It was a big undertaking, and Bombelli only managed to publish three of his planned five books, but it served to spread the learning of many of the new ideas. He developed helpful symbols for the arithmetic processes that greatly simplified their understanding. He was the first to accept and work with imaginary numbers, that is, the square roots of negative numbers. The imaginary number *i*, where $i = \sqrt{-1}$, he called *più di meno* ("plus of minus"), and –*i*, *meno di meno* ("minus of minus"). For example, he would write the complex number $3 + 2i$ as 3 p di m 2, abbreviating *più di meno*. Similarly, he would write the complex number $3 - 2i$ as 3 m di m 2. He completely illustrated the four arithmetic operations, addition, subtraction, multiplication, and division for complex numbers. This greatly simplified dealing with such numbers and how to solve for the roots of cubic and quartic equations. Bombelli's work with complex numbers was a breakthrough since, for thousands of years, such numbers, as Cardano states, were not dealt with and considered useless. Today, they play a major part in our mathematics and are of great practical use in physics and electronics. There still did not exist practical symbols in

Figure 4.9   François Viète.

algebra texts for coefficients, and there were few symbols for unknown variables. A simple formula for the quadratic equation taught today in an elementary algebra course could not be written without resorting to words or a numerical example.

The French mathematician François Viète (1540–1603; Figure 4.9) lived during a tumultuous time in France. In 1572, King Charles IX set out to massacre the Huguenots, a powerful group of Protestants of which Viète was a member. Fortunately, he was not active in their cause, and as an amateur mathematician, he made a significant contribution to algebra. The turmoil between the Catholics and the Protestants continued in France for more than a decade, and eventually Viète had to leave his position in the Court in Paris and escape to the coastal town of Beauvoir-sur-Mer. It was at this more peaceful time that he devoted himself to the study of mathematics and wrote *In Artem Analyticam isagoge* ("*Introduction to the Analytic Art*"), in which he developed the first system of algebraic notation. The book was published in 1591 (Figure 4.10), and though many of his ideas were not new, he was the first to introduce letters for all quantities, known and unknown, using vowels for the unknowns and consonants for the knowns.

In his book, Viète also improved the treatment of the theory of equations, introduced the word "coefficient", and used a fraction line to indicate division. In a second book, *Supplementum Geometriae*, published in 1593, Viète made the arduous calculation of π to 10 places using a polygon of $6 \times 2^{16} = 393{,}216$ sides, and appears to be the first to show π as an infinite sequence as $\pi = 3.1415926535$ 8979323846 2643383279 5028841971 6939937510 ... .. Though controversial, Viète is sometimes referred to as the "father of algebra." The need to attach the appellation "father" to individuals who were instrumental in changing the course of history should not diminish the importance of women's many contributions throughout the course of history.

Though algebraic contributions to mathematics during the Renaissance were the most outstanding, artist-mathematicians did make noteworthy contributions to geometry and the art of perspective. The German artist Albrecht Dürer (1471–1528; Figure 4.11) was one of eighteen children. He was an engraver who wrote a major

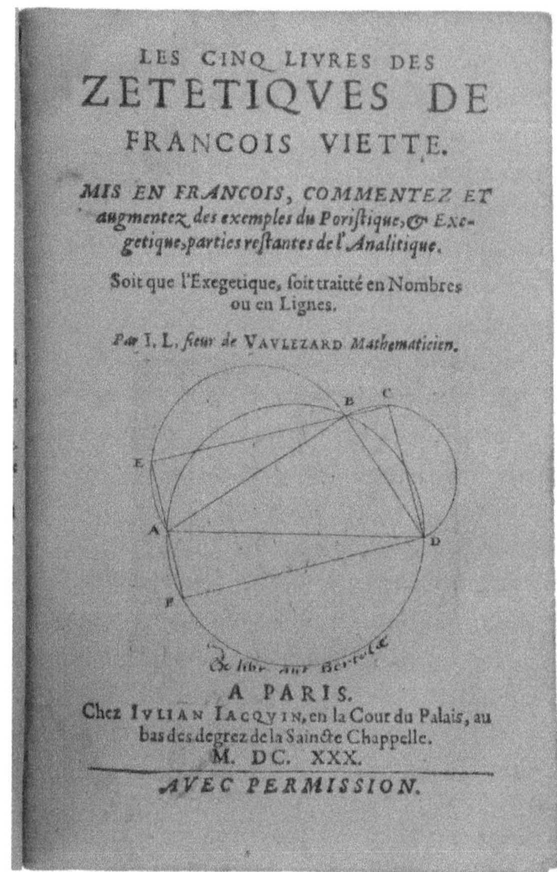

Figure 4.10   The earliest first edition (1630) of Viète's work on symbolic algebra.

work, *Unterweisung der Messung mit dem Zirkel und Richtscheit* ("*Treatise on Mensuration with the Compass and Ruler in Lines, Planes and Whole Bodies*"). This text, consisting of four books and published in 1525, was the first book on geometry and mathematics written in German, which makes Dürer one of the most influential mathematicians of the Renaissance. His mastery of geometry and the art of perspective can be observed in his series of 20 woodcuts, *Life of the Virgin*, produced between 1502 and 1511. In 1514, he created one of

Figure 4.11   Self portrait of Dürer at age 22.

his most famous engravings, *Melancholia 1*, which contains the first magic square that shrewdly placed the publication date, 1514, in the middle of the bottom row, as shown in Figure 4.12. The numbers in a magic square all add up to the same value along each row, each column, and the two main diagonals. You can check in the figure that the sum of each row, column, and diagonal is 34. This is a special magic square that Dürer created, as it possesses properties not found in other magic squares. We list a few of these intriguing special properties as follows:

The four corner numbers have a sum of 34:

$$16 + 13 + 1 + 4 = 34$$

On each corner, the four 2 by 2 squares have a sum of 34:

$$16 + 3 + 5 + 10 = 34$$

Figure 4.12   *Melancholia 1* and Dürer's magic square.

The center 2 by 2 square has a sum of 34:

$$10 + 11 + 6 + 7 = 34$$
$$2 + 13 + 11 + 8 = 34$$
$$9 + 6 + 4 + 15 = 34$$
$$7 + 12 + 14 + 1 = 34$$

Can you find some other number patterns in this beautiful magic square? Remember, this is not a typical magic square, which only requires that all the rows, columns, and main diagonals have the same sum. In the Dürer magic square, there are many more properties.

Dürer's four books contain many illustrations, including the spiral of Archimedes, regular polygons, pyramids, cylinders, the five Platonic solids, sundials, his theory of shadows, and his theory on perspective. By applying mathematics to art, Dürer produced exceptional ideas that were not only important in art but fundamental to mathematics itself.

The first German algebra book *Coss* (the "*Unknown*") was written in 1525 by the Austrian mathematician Christoff Rudolff (1499–1543) in Vienna. Algebraists were called *cossists* at this time, and algebra was considered the *cossic* art. One of his innovations was the law of exponents, where he showed that when multiplying two of the same terms with exponents, you just need to add the exponents. For example, he illustrated this with powers of 2 as follows:

$$(2^3)\,(2^4) = 2^{(4+3)} = 2^7, \quad \text{or in general: } (2^a)\,(2^b) = 2^{(a+b)}$$

As the laws of exponents became well known, it inspired the Scottish nobleman John Napier of Merchiston (1550–1617) (Figure 4.13) to construct the first table of logarithms using the new decimal fraction notation introduced by Simon Stevin (1548–1620), a Flemish mathematician. This was a prodigious task by Napier and took him 20 years to complete. Napier introduced the word logarithm, from the Greek *logos* (ratio) + *arithmos* (number). The logarithm of a

Figure 4.13   John Napier.

number when the base is 10 is the power of 10 equal to the number. For example, since $10^2 = 100$, the logarithm of 100 is 2. Similarly, the logarithm of 1000 is 3, because $10^3 = 1000$. Using logarithms greatly facilitated calculations with large numbers, which were increasingly necessary for astronomical problems. Here we present an example of how one did calculations using logarithms.

Suppose you wanted to multiply 57 by 118. You would first look up the logarithms of these numbers: log 57 = 1.75587 and log 118 = 2.07188. This means that $10^{1.75587} = 57$ and $10^{2.07188} = 118$. So, the problem in terms of powers of 10 is $(10^{1.75587}) \times (10^{2.07188})$. To do this "multiplication" problem, all you have to do is add 1.75587 + 2.07188 = 3.82775, which means the answer is then $10^{3.82775}$. To determine the actual number, one now goes to the table of logarithms and finds the number that corresponds to the logarithm 3.82775, which is 6,726. Therefore $57 \times 118 = 6,276$. Logarithms then simplify a difficult multiplication problem by reducing it to an addition problem. This calculation of course is done in a nanosecond today with a calculator or computer. However, in the 17th century it was an enormous mathematical advancement, and the use of logarithms remained in use for

400 years until digital calculators arrived in the 1970s. The author of this book had to learn how to calculate with logarithms in his second or third year of a high school mathematics course.

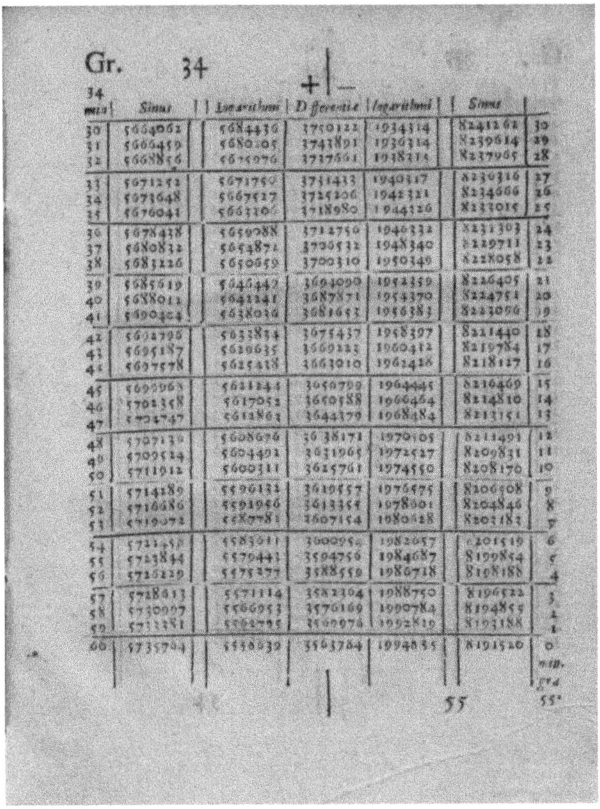

Figure 4.14   Napier's logarithms.

In 1614, Napier's logarithmic tables first appeared in a book entitled *Description of the Wonderful Canon of Logarithms* (Figure 4.14). A second work on logarithms, *Construction of the Wonderful Canon of Logarithms*, published in 1619, described the theory underlying the development of the tables. Unfortunately, these works were not easily understood and were difficult to apply. At this time, the English mathematician Henry Briggs (1561–1630) was studying eclipses and was very interested in astronomy. Astronomical calculations are quite tedious and require great accuracy, so when he saw Napier's tables of logarithms, he was so impressed with their ability to simplify his calculations that he set about to clarify their understanding. He had also published tables to aid in astronomical calculations, but Napier's tables were far more useful than his, which motivated him to redesign them. He visited Napier several times and spent many years constructing and improving the tables. After several publications of improved tables and Napier's death, Briggs published his final treatise *Arithmetica Logarithmica* in 1624. This treatise included logarithms of numbers from 1 to 20,000 and from 90,000 to 100,000 computed to 14 decimal places. An incredible achievement which was published again in 1624 in the Netherlands, and in 1633 in London, with all the missing numbers included.

During this time, there were significant advances in trigonometry because travel by sea became more extensive, and sailors needed to use trigonometry to determine their position at sea. Many texts on plane trigonometry and spherical trigonometry were written, including work done by the Greeks. The word "trigonometry" was coined by Bartholomeo Pitiscus (1561–1613), a Polish theologian, in a work published in 1600. This expansion of trigonometric knowledge enabled the remarkable discovery by the Polish mathematician and astronomer Nicolaus Copernicus (1473–1543) that the Sun was the center of the Universe, and the Earth moved around the Sun. This Heliocentric view of the solar system contradicted the age-old Geocentric view that the Earth was the center of the Universe, proposed by the Alexandrian mathematician Claudius Ptolemy (ca.100–c.170 CE). As expected, Copernicus's theory was strongly rejected by the Catholic Church and many conservative Protestants, as it contradicted biblical teachings. Copernicus found support for his theory from the brilliant Italian scientist Galileo Galilei (1564–1642), who revolutionized theories about the universe and provided the foundations for modern physics (Figure 4.15).

At the tail end of the Renaissance, Albert Girard (1595–1632), a French mathematician and musician, produced a very explicit understanding of the roots of a polynomial. In his 1629 work *Invention Nouvelle en l'Algèbre* ("*A New Discovery in Algebra*"), he provided a clear statement of what is called *The Fundamental Theorem of Algebra*.

Figure 4.15   Copernicus and Gallileo.

It applies to polynomial equations. These are equations whose terms contain powers of the variable: x, x^2, x^3, etc. In basic algebra, the simplest polynomial equation is a linear equation of the form $ax + b = 0$, which has one solution. This is followed by the general quadratic equation, $ax^2 + bx + c = 0$, which has two solutions. The next one higher is the general cubic equation, $ax^3 + bx^2 + cx + d = 0$, which has three solutions, and so on. *The Fundamental Theorem of Algebra* proves this concept that a polynomial equation has as many solutions as the highest power of the variable. This includes not just integers (positive and negative whole numbers and zero), fractions, and roots of positive numbers, but also roots of negative numbers, which are called *imaginary numbers.* We must mention here that the term "imaginary numbers" is an unfortunate one, as roots of negative numbers have applications in the real world. However, when they were first encountered, they did not appear to be useful and hence were given that name. The clear proof of *The Fundamental Theorem of Algebra* was a major contribution primarily because it accepted negative and imaginary roots, which for many years were not considered viable solutions. It is interesting to note that there is also a *Fundamental Theorem of Arithmetic*, which states that every integer greater than 1 can be uniquely factored into prime numbers. A factor is just a divisor of a number, and a prime number is a number that has no factors except itself and one. For example, the number 60 can be factored into the primes (2)(2)(3)(5). The number 60 can also be factored in other ways, but there is only one way it can be factored into primes. Albert Girard is also responsible for introducing the abbreviations. sin, cos, and tan for angles in trigonometry, and the root sign $\sqrt{\ }$, all of which are used today.

The Renaissance clearly deserves its name. Besides the enormous advancements in mathematics, science, and art that were produced, it generated a spirit which spawned a major awakening of the world. There was a departure from tradition, widespread dissemination of information, Greek works were translated into several languages, and the use of reason and the scientific method became paramount. All these factors set the foundation for the development of calculus and much of modern mathematics taught in schools today.

Chapter 5

The Foundations of Probability Theory

As the seventeenth century dawned, the Age of Enlightenment emerged from the Renaissance, promoting religious tolerance, civilizing reason, cherishing individual liberty, and honoring nature. This view of the world spread throughout Western Europe. Mathematics, science, art, philosophy, and politics continued to advance and had a marked effect on people's lives. In 1637, the French philosopher, scientist, and seminal mathematician René Descartes (1596–1650) (Figure 5.1) published his essay *Discours de la Méthode* (*"Discourse on the Method"*) at Leiden in the Netherlands. This essay contains his famous maxim: *Cogito ergo sum* ("I think therefore I am").

Figure 5.1   René Descartes.

Some attribute the beginning of the Enlightenment to Descartes' essay, which he believed contained a better way to acquire knowledge than by applying Aristotle's logic. Descartes claimed that mathematics is certain, so all knowledge must be based on mathematics.

Descartes is most well-known for applying algebra to geometry in his book *La Géométrie*, published in 1637, creating analytic geometry, which is called in secondary schools today coordinate or Cartesian geometry. It consists of a graph with a horizontal, or *x* axis, and a vertical, or *y* axis, where each point is assigned two coordinates, *x* and *y*, that represent the distance of the point from each axis (Figure 5.2). He changed the letter convention in use at the time, employing letters near the end of the alphabet, *x*, *y*, and *z*, for the unknowns, and letters near the beginning of the alphabet, *a*, *b*, and *c*, for the knowns, which is still the practice used today. He was also responsible for introducing the present method of using superscripts for powers or exponents, such as the 3 in the expression x^3 to indicate raising x to

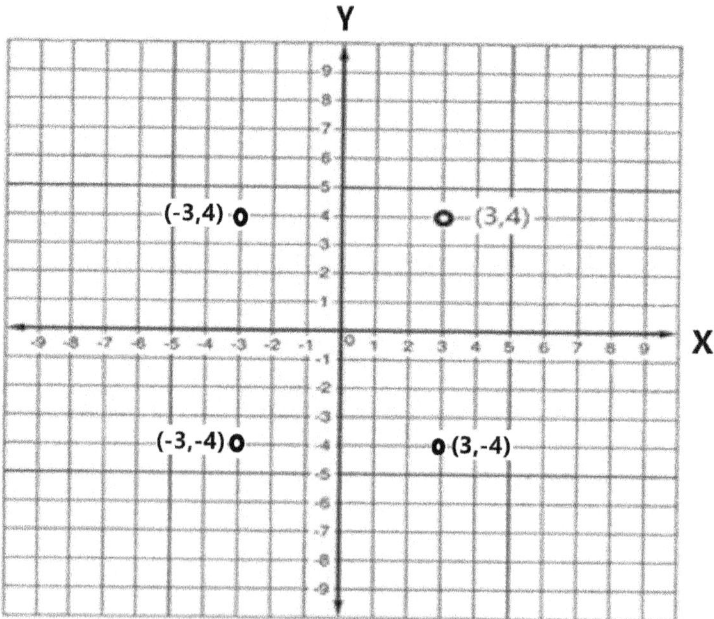

Figure 5.2   Cartesian graph.

the third power. Descartes' work is considered by many to have had a major influence on the development of calculus.

In 1647, when Descartes was 51 and one of the most renowned mathematicians in Europe, he asked to meet with the prominent French mathematician and philosopher Blaise Pascal (1623–1662; Figure 5.3). Pascal, like Descartes, was destined to become one of the most influential mathematicians of the seventeenth century. Descartes held that you were either a "Cartesian" who embraced his philosophy, or you were not. Young Pascal at 24 was not. Pascal was skeptical about Descartes' claim that one could reason the existence of God and rejected his idea that a vacuum could not exist in nature, having shown through experimentation that it could exist. Pascal showed Descartes a calculating machine he had invented that could add and subtract. This was one of the first of its kind (Figure 5.4). The visit and discussion did not go well. Descartes took his leave and wrote in a letter to a colleague claiming that Pascal *"has too much vacuum in his head."*

However, he could not help but hold Pascal in esteem and returned the next morning to care for him, as he was ill, and helped him to get

Figure 5.3   Blaise Pascal and a commemorative statue (Louvre Collections).

Figure 5.4   The Pascaline, one of the first digital calculators, created by Pascal.

better. Remembering Descartes' concern, Pascal later commented: *"The heart has its reasons which reason knows nothing of."*

Blaise Pascal's mother died when he was only three years old, and death in those days was unfortunately a very common experience. Though his father was a learned judge who enjoyed science and mathematics, for some reason, he did not want his son to study mathematics until he was 15. He decided to self-teach him and removed all mathematics texts from the house.

However, Pascal, being the prodigy that he was, discovered geometry when he was 12, and was able to determine that the sum of the angles of a triangle adds up to 180 degrees or two right angles. His father, upon witnessing his son's ability, decided to give him a copy of Euclid's Elements, which he devoured. At the age of 16, Pascal proved several theorems in projective geometry, a branch of mathematics which deals with projections of figures onto another surface such as a shadow cast by an object, or the mapping of the earth onto a flat surface which are examples of projective geometry. At age 17, Pascal published his first work, *Essai sur les Sections Coniques* (*"Essay on Conic Sections"*). Conic sections are the curves formed when a plane intersects a cone at various angles. Four different curves are created in this way: the circle, the parabola, the ellipse, and the hyperbola, as shown in Figure 5.5.

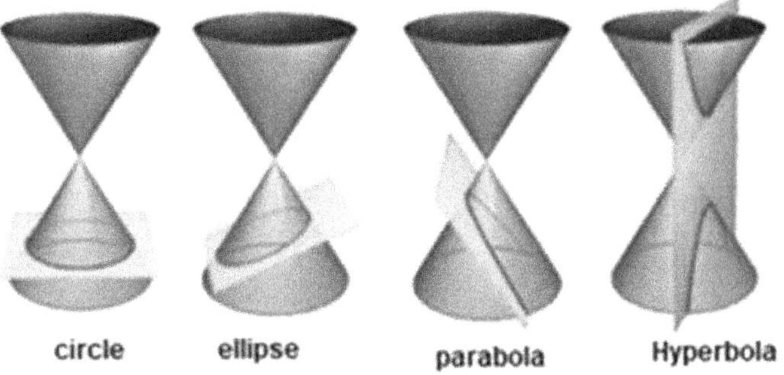

Figure 5.5   Four conic sections.

René Descartes, upon reading Pascal's essay, became skeptical and assumed that his father wrote it. When he discovered otherwise, he gave little credit to Pascal and stated, *"matters related to this subject can be proposed that would scarcely occur to a 16-year-old child."*[1]

Pascal started constructing his calculator when he was 19, finishing it three years later to help his father, who had become the supervisor of taxes in the French city of Rouen and had to do tedious calculations. His calculators were not easy to operate and were expensive, which limited their production and led to their eventual discontinuation. Still, they were a notable achievement and inspired the development of more practical and useful designs. Several of his calculators still survive and are in museums in Paris, France, and Dresden, Germany.

When Pascal was only 23, his father fell on an icy street and broke his hip. Fortunately, he received good medical care and managed to walk again, but died five years later, leaving Pascal very distraught, and he began to embrace the Christian religion. He devoted his life to mathematics and science, publishing additional work on conic sections in 1648. In 1654, he wrote *Traité du Triangle Arithmétique* (*"Treatise on the Arithmetical Triangle"*), which was published

---

[1] *The Story of Civilization: Volume 8, The Age of Louis XIV* by Will & Ariel Durant; Simon and Schuster, New York, 1963, chapter II, subsection 4.1 p. 56.

posthumously in 1665. In this treatise on the triangle, he presents an important and convenient way to show the binomial coefficients in a triangular form. Though Pascal was not the first to study the binomial coefficients, his clear presentation honors him by naming it *Pascal's Triangle*. (See Figure 4.3 in Chapter 4 for more information on Pascal's triangle.)

He also developed and used the logical method of proof called *mathematical induction,* which is a clever and effective method of proof that begins by assuming a true relationship between a set of numbers or mathematical expressions and shows that this relationship is true for all possible subsequent sets of numbers or expressions. It could be considered analogous to a succession of standing dominoes, and when one tips, the first one triggers all the others to fall one after another. It differs from the more standard form of proof used by the ancient Greeks, which is mathematical deduction, which starts with accepted axioms or previously proven theorems and logically develops the proof of a new statement or theorem.

Here is a clear example of how mathematical induction works. We will prove that the square of any odd number is also an odd number. We know that if we start with the first odd number, 1, and square it, we get an odd number, $1^2 = 1$. We now consider *any* odd number n and see what happens if we assume n^2 is also an odd number. This may not be true, but observe what we do with this assumption. If n^2 is an odd number, let us see what is true for the next odd number $n + 2$ when we square it. Starting with $(n + 2)^2$, we expand this product of binomials by multiplying out the four possible products and simplifying the result. You may remember how this works from a basic algebra course. It is called the **FOIL** method, which stands for **F**irst product, **O**utside product, **I**nside product, and **L**ast product:

$$\mathbf{F} \qquad \mathbf{O} \qquad \mathbf{I} \qquad \mathbf{L}$$

$$(n + 2)(n + 2) = (n)(n) + (n)(2) + (2)(n) + (2)(2) = n^2 + 2n + 2n + 4 = n^2 + 4n + 4$$

Let us examine the result $n^2 + 4n + 4$. The last two terms must be even numbers because they can be divided by 4. The first term, however, is an odd number because we assumed that n^2 is an odd number. Therefore, the result, which is the sum of an even number and an odd number, will be an odd number. The principle of mathematical induction now reasons this way. We have shown that for any odd number n, where n^2 is odd, the following odd number, $n + 2$, when squared, is also an odd number. Therefore, because the square of the first odd number 1 is indeed odd, and we have proven that the square of the next odd number is also odd, it must be true for the next odd number 3: $3^2 = 9$. If we now apply the proof to the odd number 3, it must be true that the square of the next odd number 5 is also odd. We can continue to apply this result to all the subsequent odd numbers, and therefore, it must be true for all odd numbers. Hence the metaphor of all the dominoes falling as a result of the first one tipping over.

One of the oldest human activities is trying to determine the outcome of events, such as: What will the weather be like? Will I like my new job? Which team will win the game? What is the chance of winning money in the lottery or the casino? In each situation, it helps to understand the odds of a favorable or unfavorable outcome. Such was the dilemma of the French writer Antoine Gombaud, alias Chevalier de Méré (1607–1684). He was not a true nobleman, but he adopted the title *chevalier* (knight) to reflect the characters he identified with in his dialogues, and *Méré*, given his education in Méré, France. He was an extraordinary gambler and an amateur mathematician who pondered the age-old question of how to decide the outcome of a game of chance if it is not yet completed. Consider two players who agree to play five games of chance, with each putting up the same amount of money, and whoever wins three games is the overall winner. Now, suppose the play is interrupted after three games, with one player having won two games and the other player having won one game. The problem is, then, how to fairly divide the money between them. This problem, and another one that Gombaud grappled with, was the chance of getting at least one pair of sixes when throwing a pair of dice 24 times, where the six sides of each die contain the numbers 1 to 6. Dice, it turns out, have been discovered in many ancient cultures, but the nature of many of the games is not known. Gombaud reasoned that there should be more than a 50% chance of two sixes showing up in 24 tosses, and therefore, if he were to bet even money on this outcome, it should be profitable. However, this was not his experience, and so he proposed these two problems to the Mersenne Salon, which was a salon organized by a French monk, Marin Mersenne (1588–1648), who arranged correspondence between noted scientists and philosophers.

Blaise Pascal and another distinguished French mathematician, Pierre de Fermat (1601–1665) (Figure 5.6), decided to take on these two stimulating challenges. Though both these problems had been studied before by Gerolamo Cardano, who was also an avid gambler, he produced no definitive results. During the summer of 1654, Pascal and Fermat corresponded in writing with several letters and began to develop the foundation for the present theory of probability. They solved Gombaud's division problem for two players, but they were not able to establish the mathematical solution for three or more players, which is far more challenging than for two players. We describe one of the solutions here to give the reader an idea of what needs to be considered. The assumption is that each player contributes the

Figure 5.6   Pierre de Fermat and his statue in the Museé des Augustins.

same amount of money to the game, and the winner of three games is entitled to all the money. Pascal began by saying that if both players are equally matched and the play is interrupted, the money should be equally divided. In other words, the particular outcome of the game should not count. On the other hand, if the players agree that if the game is not completed, each should receive an equitable amount based on their scores, then it is necessary to formulate fair rules of division. Pascal states that the amount each should receive when the play stops should be determined by how many games remain to be played. For example, suppose that the total amount of money is $60. If, when the play stops, the number of games that need to be won is the same for each player to be the winner, then clearly the stake should be divided evenly, $30 for each player. If, however, when the game stops, player 1 needs one game to win and player 2 needs two games to win, then Pascal considers all the possible outcomes and reasons as follows. Suppose if one more game were played and player 2 were to win it. The scores would then be even, and each player should receive half of the stake, $30. On the other hand, suppose player 1 were to win one more game, then he would receive all the money, $60. So, if you take the average of the two possible outcomes for player 1, $30 and $60, you come up with $45, which is what player 1 should be fairly entitled to when the game is stopped. This example is a small part of the entire solution, which is mathematically more complex.

In the other challenge that Gombaud proposed, involving the chance of getting two sixes in 24 tosses of two dice, he correctly reasoned that since there are six faces on each die, there are 36 combinations that can appear on the two dice. Since only one of these combinations is two sixes, the chance of throwing two sixes is $\frac{1}{36}$. He then incorrectly calculated the chance of success by multiplying $24 \times \frac{1}{36} = \frac{2}{3} = 0.6667$, causing him to bet a lot of money on this outcome. Pascal calculated the actual probability of getting at least one double six in 24 throws to be 0.4914, or a little less than 50%. This explains why Gombaud's odds were against him. This probability is correctly determined as follows. One first considers the chance of not getting a double six on one toss, which is $\frac{35}{36}$. Now the probability of *not* getting a double six on 24 tosses is determined by multiplying this fraction 24 times:

$$\left(\frac{35}{36}\right)^{24} = 0.5086$$

It is assumed that the probability of getting a favorable outcome added to the probability of not getting a favorable outcome is 100% or 1.00. This is because it represents all the possible outcomes. We can then subtract the above result, 0.5086, from 1.00 to find the probability of actually getting a double six in 24 tosses:

$$1 - \left(\frac{35}{36}\right)^{24} = 1 - 0.5086 = 0.4914$$

Antoine Gombaud must have thought highly of himself because he believed that he had discovered probability theory, and that based on his probability calculations, mathematics was inconsistent. Leading mathematicians have given none of his convictions any merit.

Probability theory, and its sister subject, statistics, have progressed over the years into a large body of very useful mathematical concepts, some of which we are exposed to every day, such as the weather forecast, the lottery, or election predictions. There are two types of probability, theoretical and experimental. Theoretical probability is computed by assuming logical formulas and using available facts and information. For example, if a coin is perfectly balanced and tossed randomly, it is assumed that the *theoretical probability* of it landing heads is $\frac{1}{2}$ or 50%. On the other hand, if one tosses an actual coin many times, at least 1000, and it lands heads 550 times, then we say that the *experimental probability* is $\frac{550}{1000}$ or 55%. Now the weather forecast is more complex when presented in terms of the theoretical probability of something happening, which takes into consideration many atmospheric factors. A probability of 100% means an event will happen, and 0% that it will not. But how should one interpret what is meant by the statement "There is a 50% chance of rain"? It does not mean that 50% of the time it will rain in the forecast area or that 50% of the forecast area will experience rain. At best, it means that any part of the forecast area has a 50% chance of precipitation, and that the amount of rain will be at least 0.010 inches. So clearly, probability is anything but an exact science, and there is always risk associated with any prediction.

Throughout his life, Pascal suffered from chronically poor health, especially during his correspondence with Fermat, and was bedridden quite often. However, despite his illness, he worked diligently on scientific and mathematical problems, making significant contributions in the fields of fluid dynamics and atmospheric pressure. In 1654, he claimed to have had a strong religious experience, which caused him to devote his life more intensely to Christianity. He proposed a wager about God, applying his ideas of probability. Pascal reasoned that since we have no proof of the existence of God, one should consider the four possible outcomes based on whether one chooses to believe, or not believe, in the existence of God as follows:

"Gamble"	God does exist	God does not exist
Believe in God and live a reverent life	Infinite Gain	No Gain or Loss
Not believe in God and live a hedonistic life	Infinite Loss	Finite Gain

Clearly, for Pascal, if you accept his premise, one has much more to gain by believing in God, and for him, it was the only choice. Pascal's greatest contribution was the foundations of probability, but sadly, like many great people at that time, his life was cut short at age 39.

Fermat, along with Descartes and Pascal, was also one of the world's foremost mathematicians of the seventeenth century. He was fluent in six languages: French, Latin, Occitan, classical Greek, Italian, and Spanish. In 1631, he became a lawyer and a government official in Toulouse, France, which entitled him to change his original name from Pierre Fermat to Pierre de Fermat. The plague, most likely bubonic plague, a devastating disease, struck this part of France in 1650, and many succumbed to the disease within a few weeks. In 1653, Fermat contracted it, but he was quite bemused, as he did not report his situation and was thought to be dead; however, it was soon corrected by a colleague who observed:

> *I informed you earlier of the death of Fermat. He is alive, and we no longer fear for his health, even though we had counted him among the dead a short time ago.*

Fermat loved mathematics, and for him it was more of an avocation than a profession. He made important contributions to analytical geometry, probability, number theory, and calculus. In 1637, he completed a manuscript on analytic geometry, which independently contained many of the ideas put forth by Descartes. He was also intrigued by number theory and studied prime numbers and perfect numbers, so named by the Greeks. A *perfect number* is equal to the sum of its divisors. The smallest perfect number is $6 = 1 + 2 + 3$ and the next larger perfect number is $28 = 1 + 2 + 4 + 7 + 14$. The Greeks had only discovered four perfect numbers: 6, 28, 496, and 8128, so Fermat was motivated to find a formula to reveal other perfect numbers. He is well known for what is called today *Fermat's Little Theorem*, which is an important result that has many applications and helps identify prime numbers and perfect numbers. It states that if you take any prime number p and any integer a, then the prime number p will be a divisor of the number $a^p - a$. Here is a clear example of Fermat's Little Theorem. If p equals the prime number 3, and a equals the integer 5, then $a^p - a = (5)^3 - 5 = 125 - 5 = 120$. This tells us that the prime number 3 must be a divisor of 120.

Fermat kept his methods secret, and though he wrote about his results with pleasure to his colleagues, there is scarcely any record of his proofs or procedures. It was not until the middle of the eighteenth century that many of his contributions became more widely known. The most famous example of this is known as *Fermat's Last Theorem*. It is related to the famous Pythagorean theorem, which you probably recall from a basic geometry course. The Pythagorean theorem states that for a right triangle whose sides are lengths a and b and whose hypotenuse is length c, we have the relationship: $\boldsymbol{a^2 + b^2 = c^2}$

(Figure 5.7). Now one can find many sets of three integers that satisfy this theorem, such as:

$$3^2 + 4^2 = 5^2 \quad \text{or} \quad 9 + 16 = 25$$

$$5^2 + 12^2 = 13^2 \quad \text{or} \quad 25 + 144 = 169$$

$$8^2 + 15^2 = 17^2 \quad \text{or} \quad 64 + 225 = 289$$

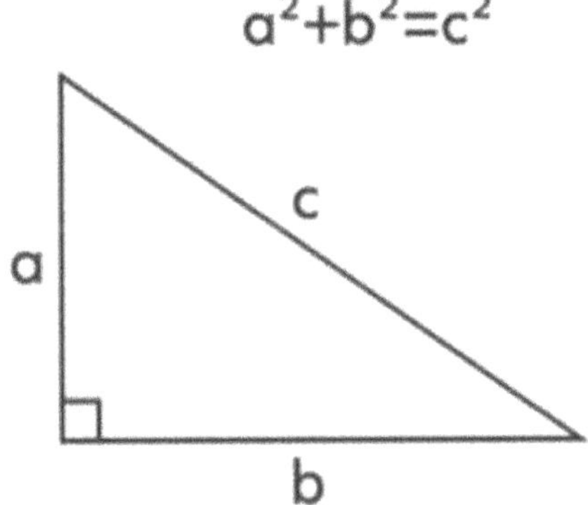

Figure 5.7   Pythagorean theorem.

Actually there are an infinite number of sets of three integers that can be found to satisfy the Pythagorean theorem, and they can be generated using the following formula developed by Euclid. A set of three numbers that satisfies the Pythagorean theorem is called a Pythagorean triple. To generate a Pythagorean triple, you choose any two integers m and n, where $m > n > 0$, and calculate as follows: $a = m^2 - n^2$, $b = 2mn$, and $c = m^2 + n^2$. For example, if $m = 7$ and $n = 4$:

$$a = 7^2 - 4^2 = 49 - 16 = 33,$$

$$b = 2(7)(4) = 56, \text{ and}$$

$$c = 7^2 + 4^2 = 49 + 16 = 65.$$

It is then true that $33^2 + 56^2 = 65^2$, or $1089 + 3136 = 4225$.

What Fermat stated in his Last Theorem is that it is impossible to find three integers a, b, and c such that $a^4 + b^4 = c^4$, which he indeed proved. In addition, he made a more general statement that it is impossible to find three integers a, b, and c such that $a^n + b^n = c^n$, where n is *any* integer larger than 2. This is quite surprising and is a cause for wonder as to why it should work for squares and not for cubes or any other higher power. Mathematical curiosities like this one have always intrigued mathematicians, especially those who are motivated to study relationships involving numbers, whether or not their results have any practical value. The field is called Number Theory, and it prompted the famous English mathematician G.H. Hardy (1877–1947), who devoted much of his life to the study of number patterns, to write *A Mathematician's Apology* in 1940, in which he states:

> *The mathematician's patterns, like those of the painter's or the poet's, must be beautiful, the ideas, like the colours or the words, must fit together in a harmonious way. There is no permanent place in the world for ugly mathematics.*

Hardy's book has inspired many students to appreciate mathematics purely for its beauty and to be motivated to pursue its study.

Now Fermat, who was known for rarely putting his work in a clear, organized form, claimed his famous Last Theorem in the margin of his copy of a work by the Greek Diophantus, to which he attached this statement:

> *I have discovered a truly remarkable proof which this margin is too small to contain.*

His son Samuel was the first to discover Fermat's Last Theorem, and he published it along with Fermat's note in 1670. Mathematicians were skeptical that he had indeed found a proof, as it was an extremely challenging problem and actually remained unproven for over 300 years. Sir Andrew Wiles proved it in 1994 by applying advanced modern methods that were not available to Fermat. It is still to

Fermat's credit that he was able to somehow envision that this curious property of the powers of integers was true.

Both Blaise Pascal and Pierre de Fermat were very gifted mathematicians, and their contributions to analytic geometry, probability, and number theory laid some of the groundwork for the discoveries of calculus by Isaac Newton (1642–1726) and Gottfried Leibniz (1646–1716) during the seventeenth century. Few great discoveries can be fully accredited to one person and they emerge as the culmination of many prior contributions.

Chapter 6

The Discovery of Calculus

Calculus — the very word conjures up an image to many people of a branch of mathematics that is both advanced and challenging, yet there is generally a desire to know what it is all about. This chapter will not only reveal the extensive history and the genius that created this seminal branch of mathematics, but it will also clearly explain the underlying concepts so that you can get a clear understanding of what calculus is all about. The elusive ideas of infinity and infinitesimal are what calculus grapples with and succeeds in mastering. Its origins go back to the Greeks, particularly Zeno, Eudoxus, and Archimedes. As described in Chapter 2, the Greek philosopher Zeno of Elea (495–430 BCE) was one of the first to grapple with the concept of infinity. He illustrates in three different paradoxes that the idea of motion can be viewed in such a way that it appears to be impossible. The runner in Chapter 2 attempting to run a distance must first complete half the distance, then a quarter of the distance, then an eighth of the distance, then a sixteenth of the distance, and so on. If one adds up all the fractional distances that comprise the total distance, the sum goes on forever, and therefore it appears that the runner can never complete the total distance. Expanding this argument, one can conclude that if the runner cannot complete this distance, then the runner cannot complete any distance, and thus, one concludes that motion is impossible. This is equivalent to making sense of the sum of the infinite series:

$$\frac{1}{2}+\frac{1}{4}+\frac{1}{8}+\frac{1}{16}+\cdots$$

Calculus simply solves this argument by first observing that the sum of these fractions gets infinitesimally closer and closer to the number 1 but never exceeds 1. Therefore, calculus essentially "defines" the infinite sum to be the number 1 by employing the word limit (abbreviated Lim) and states:

$$\mathbf{Lim}\left(\frac{1}{2}+\frac{1}{4}+\frac{1}{8}+\frac{1}{16}+\cdots\right)=1$$

That basically eliminates the paradox. This type of reasoning is found in the branch of calculus called *Differential Calculus*, and the above limit is expressed mathematically as follows:

$$\sum_{n=1}^{n=\infty}\left(\frac{1}{2^n}\right)=1$$

The capital Greek letter sigma Σ stands for the sum of the fractions $\left(\frac{1}{2^n}\right)$, and they are added from $n=1$ to $n=\infty$ (infinity).

The classical Greek mathematician Eudoxus of Cnidus (390–340 BCE) was one of the first to develop concepts that are fundamental to the other branch of calculus called *Integral Calculus*. His *Method of Exhaustion* was applied by Archimedes (287–212 BCE) to find the circumference of a circle by both inscribing and circumscribing the circle with polygons containing a large number of sides. In this way, the average of the perimeters of the polygons approximated the circumference of the circle and allowed Archimedes to accurately calculate the circumference. Then, knowing the diameter of the circle and using the fact that π is the circumference divided by the diameter, he determined a range of values for π between $3\frac{10}{71}$ and $3\frac{1}{7}$. (See Figure 6.1.)[1]

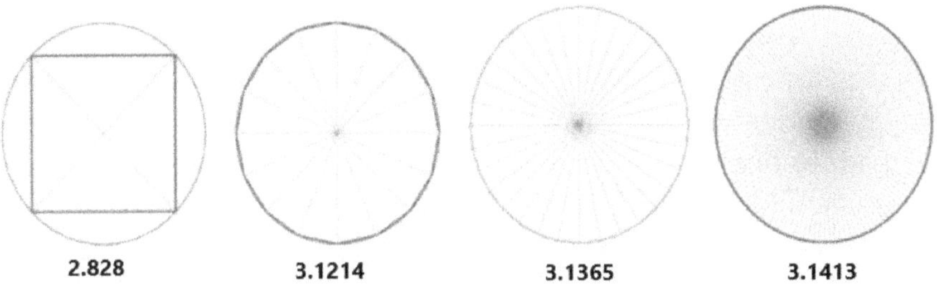

Figure 6.1   Inscribing polygons to approximate π.

We therefore have two branches of calculus. Differential calculus deals with the instantaneous rate of change of a variable, such as distance or velocity while Integral calculus deals with the areas enclosed by curves and geometric figures such as circles and parabolas. These areas can represent physical phenomena such as force or electrical power. Though it is not immediately clear, the two branches of calculus are actually mathematical inverses of each other like subtraction is the inverse of addition, and division is the inverse of multiplication.

These mathematical innovations by the ancient Greeks did not progress further until almost a thousand years later, with the Renaissance and the advent of analytic geometry by Descartes. Many mathematicians and scientists over the years worked on similar problems dealing with infinite and infinitesimal quantities, but their methods required intricate calculations and were not easy to work with. Descartes' analytic geometry allowed curves to be represented as algebraic equations, enabling formulas to be found that expressed

---

[1] See Chapter 2 for more information.

their tangent lines, and calculated areas enclosed by curves. The tangent line to a curve is important because it shows at a point the direction of the curve and its rate of increase or decrease, measured by the angle of the tangent line.

Fermat made one of the major contributions during the early seventeenth century that was instrumental in the discovery of differential calculus. He studied the points on a curve that are local maximum and minimum values, so-called extrema, and formulated analytic methods to find these points. At a local maximum or minimum point, the tangent is horizontal, which is the direction of the curve, so it is neither increasing nor decreasing. (See Figure 6.2.) Therefore, these are important points because the rate of change of the curve at that point is zero. For example, if the curve represented the velocity of an object, the local maximum or minimum point would indicate when the velocity is zero and the object is not moving at that instant.

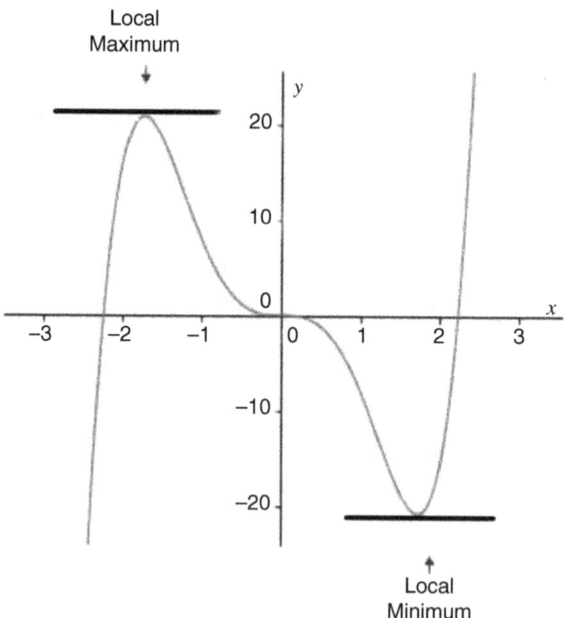

Figure 6.2   Fermat's extrema.

Descartes also came up with a method for finding tangents to a curve, but Fermat's method is the one still used today in differential calculus.

Fermat and other mathematicians applied the method of exhaustion and similar procedures to find areas bounded by curves. They were able to find formulas for areas under parabolas, hyperbolas, and sine curves. Figure 6.3 shows that the area under the parabola described by the equation $y - x^2$ between $x = 0$ and $x = 1$, is exactly $\frac{1}{3}$ units. The area under the sine curve, $y = \sin x$, between $x = 0$ and $x = \pi$ is exactly 2 units. We show this results from calculus because it is somewhat surprising that these areas are simple rational numbers.

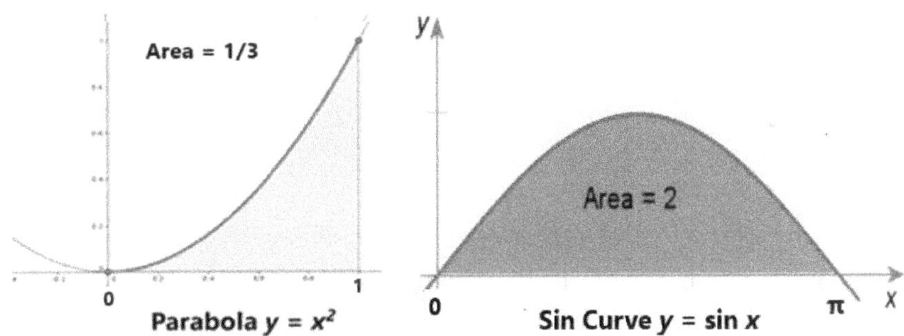

Figure 6.3   Areas under curves.

The universal geniuses Sir Isaac Newton (1643–1727; Figure 6.4) in England and Gottfried Wilhelm Leibniz (1646–1716) in Germany, working independently, brought together all the work done with tangents to curves and areas bounded by curves and created the cohesive structure of calculus. The word calculus was used before Newton and Leibniz used it for their findings, as it was previously referred to any aspect of mathematics. "Calculus" means "small stone" in Latin, and since stones were used as an aid in measuring distances, the term was applied to this new method of computation. Calculus is the foundation for modern mathematical analysis and is an essential tool in all the sciences, economics, and the social sciences. Here is the fascinating story behind those two great men and their discoveries.

The year 1665 was a tragic one in England, as it was throughout western Europe. The Great Plague of London in 1665–1666 struck without mercy. Those afflicted had a 30% chance of dying within two weeks. It killed almost 100,000 people, about one-fifth of the population of London. Yet something significant did emerge during this time. Throughout history, there have been times when a person emerges and leaves a legacy that changes the world forever. Such a person was Isaac Newton, who, when the Plague struck London, was a 22-year-old student who had just received his bachelor's degree from Trinity College at the University of Cambridge. Fortunately for him (and for us), the University closed, and he left London, and sequestered for two years in the county of Lincolnshire, where he embarked on one of the greatest mathematical and scientific journeys of all time. What emerged was not just the discovery of calculus, but the field of Newtonian physics encompassing the Laws of Motion and Gravity, a large part of which relies on the foundations of calculus. He also produced a treatise on optics.

Figure 6.4   Sir Isaac Newton.

Newton's early life was not exactly auspicious. He began his journey into the world prematurely, his father dying three months before he was born, and he was so tiny that his mother claimed he could fit inside a quart container. He had to wear a support around his neck to hold up his head, as he was very frail. His mother soon remarried and had three more children, leaving Isaac to be cared for by his grandmother. He loathed his stepfather and felt bitter towards his mother, expressing thoughts as a teenager of burning down their house. Coming from this background, he was slow to develop a passion for learning, but eventually, at the age of 17, he was taken under the care of the headmaster of the Free Grammar School in Grantham, England, and began to show his talent and embrace his education. There are various anecdotes about him displaying his abilities at this time, but no clear picture emerges of any exceptional skill.

His uncle encouraged Newton to enter Trinity College at Cambridge in 1661, although he was older than many of his fellow students. Though he accepted a position to serve other students, which brought with it an allowance toward college expenses, he seems to have preferred associating with an upper class of students. His mother had acquired some extra means at this time and most likely helped him with his finances. It was in his later years at Trinity that Newton's genius was fortunately recognized by Isaac Barrow (1630–1677), a distinguished professor of mathematics, who steered him to the study of optics and at times collaborated with him. Newton now began to devour the Greek philosophy of Aristotle and Plato, Euclid's *Elements*, Descartes' geometry, Galileo's astronomy, and many of the scientific and mathematical works of the seventeenth century. Though Newton earned the title of scholar in 1664 and received his bachelor's degree in 1665, his genius was yet to emerge. The plague may have fostered his brilliance because in the span of the next two years, when he was sequestered at Lincolnshire, England, he produced extraordinary advances in mathematics, physics, astronomy, and optics, while still not yet 25 years old. He is most notably remembered for his Laws of Gravity and his Laws of Motion, which stem from the foundations of calculus:

Newton's Laws of Motion
Law of Inertia: An object at rest tends to remain at rest.

              An object in motion tends to remain in motion.

Law of Force: The acceleration of an object of constant mass is proportional to the force applied.
Law of Action & Reaction: When a force is exerted on an object, the object exerts an equal and opposite force.

Though there are exceptions, great achievements tend to be the product of young minds, possibly because one's mind is sharp and not constrained by past ideas and beliefs, and one can see the world in a different light. Isaac Newton, Archimedes, Albert Einstein, Fibonacci, Wolfgang Amadeus Mozart, Blaise Pascal, and Pablo Picasso, to name a few, all made significant contributions before the age of 30. Newton never surpassed the achievements that he made in those two years, and in his seventy-third year had this to say:

In those days I was in the prime of my age for invention, and minded mathematics and philosophy more than at any time since.

To better grasp some of the basic concepts of calculus, we present a helpful example of the role that calculus plays in the Law of Gravity. Differential calculus is the mathematics of change. It enables one to calculate the instantaneous change of a variable relative to the instantaneous change of another variable. For example, the velocity of an accelerating object at an instant is the instantaneous change in distance relative to the instantaneous change in time. An instant of time can be thought of as a nanosecond, which is a billionth of a second. However, calculus strives to determine the instantaneous velocity of a falling object over a period that is theoretically zero seconds. The speedometer in a car does essentially that. It shows the velocity at every instant. A thermometer also displays the temperature at an instant of time. Another example is the rate of a chemical reaction relative to the instantaneous change in the temperature, or the instantaneous change of the force on an object relative to the instantaneous change in the direction of the force. Consider now an object falling, due to the gravitational force, whose velocity is constantly increasing. We have the following relation between the distance d in feet that the object falls, as a function of the time t in seconds:

$$d = 16t^2$$

This formula enables us to calculate the change in the distance over a *finite* interval of time, which is the *average velocity* over that time interval. Consider the average velocity over the time interval from $t = 0$ sec to $t = 1$ sec. The change in time is $1 - 0 = 1$ second. The change in the distance can be calculated from the above formula by taking the difference in the value of d from 1 second to 0 seconds:

$$\text{Change in distance} = 16(1)^2 - 16(0)^2 = 16 \text{ ft}$$

The average velocity, v_{avg}, from 0 seconds to 1 second is then:

$$v_{avg} = \frac{\textit{Change in distance}}{\textit{Change in time}} = \frac{16}{1} = 16 \text{ ft/sec}$$

Suppose we wanted to know the instantaneous velocity at *exactly* one second. This is where calculus comes in. In differential calculus, we accomplish this by continually decreasing both the time interval and the change in the distance, so that both differences approach zero. We then observe what happens to the ratio of the change in the distance divided by the time interval. If it tends to get closer and closer to a certain value and not to exceed that value, we assume that value is the instantaneous velocity. Table 6.1 below shows the series of calculations. We start with a time interval of 0.5 sec, and then decrease it to 0.1 sec, to 0.01 sec, to 0.001 sec, and finally to 0.0001 sec.

We calculate the distance covered over these increasingly smaller time intervals by applying the formula: $d = 16t^2$. Then we calculate the average velocity over these smaller time intervals by dividing the change in the distance by the time interval.

The last calculation gives the average velocity over the last ten-thousandth of a second before exactly one second. You can now observe in Table 6.1 that the average velocity approaches 32 ft/sec. Suppose we continue to do this with even smaller time intervals. In that case, we can prove using calculus that the average velocity gets infinitesimally close to 32 ft/sec and will not exceed 32 ft/sec. We can therefore correctly deduce that the instantaneous velocity at exactly 1 second is 32 ft/sec. In this example, we are dealing with both the infinite and the infinitesimal. The time interval, theoretically, is reduced an infinite number of times, while the size of the intervals becomes infinitesimal. We call the instantaneous velocity the *derivative* of the distance, and the process of finding a derivative is called differentiation. There is a simple procedure for finding the derivative of $d = 16t^2$, and it yields the formula $v = 32t$. This formula gives the instantaneous velocity v at any instant of time, where 32 ft/sec every second is the acceleration of gravity. Using this formula, the instantaneous velocity at 1 sec is then $v = 32(1) = 32$ ft/sec, which is the result arrived at from Table 6.1. Similarly, the instantaneous velocity at 2 sec is $v = 32(2) = 64$ ft/sec and at 3 sec, $v = 32(3) = 96$ ft/sec. This process greatly simplifies what was a difficult problem for mathematicians and physicists for many years. Differential calculus produces derivative formulas for almost any function and enables us to find instantaneous changes of any variable with respect to any other variable. For example, suppose you want the instantaneous change in the temperature of a body of water with respect to the amount of heat being added. We start with the formula or function relating the temperature to the heat added and then apply the procedure for finding the derivative of this function. This produces another function that can give you the rate of change of the temperature for any value of the heat being added.

Issac Newton formulated these ideas, which he called the 'Method of Fluxions,' over two years and returned to Cambridge in 1667 when the University reopened. It was his incredible insight that led to the discovery of differentiation, and also the fact that it was the

Table 6.1   Average velocities close to 1 second.

Time Interval from 1.0 sec	Change in Distance from 16 ft	Average Velocity = Change in Distance/Time Interval
$1 - 0.5 = 0.5$	$16 - 16(0.5)^2 = 12$	$12/0.5 = 24$ ft/sec
$1 - 0.9 = 0.1$	$16 - 16(0.9)^2 = 3.04$	$3.04/0.1 = 30.4$ ft/sec
$1 - 0.99 = 0.01$	$16 - 16(0.99)^2 = 0.3184$	$0.3184/0.01 = 31.84$ ft/sec
$1 - 0.999 = 0.001$	$16 - 16(0.999)^2 = 0.031984$	$0.031984/0.001 = 31.984$ ft/sec
$1 - 0.9999 = 0.0001$	$16 - 16(0.9999)^2 = 0.00319984$	$0.00319984/0.0001 = 31.9984$ ft/sec

mathematical inverse of integrating a function, which represents an area enclosed by the function. Years later, in 1671, he wrote the treatise *De Methodis Serierum et Fluxionum* ("*On the Methods of Series and Flows*"), containing these ideas. Still, he was hesitant to publish his calculus because he feared controversy and criticism. It was not until 1736 that it appeared in print when John Colson produced an English translation (Figure 6.5). Newton was able to show that many different existing procedures used to solve unrelated problems, such as finding maxima and minima of functions, calculating tangents to curves, and determining areas bounded by curves, could be more simply solved with his new analytical methods.

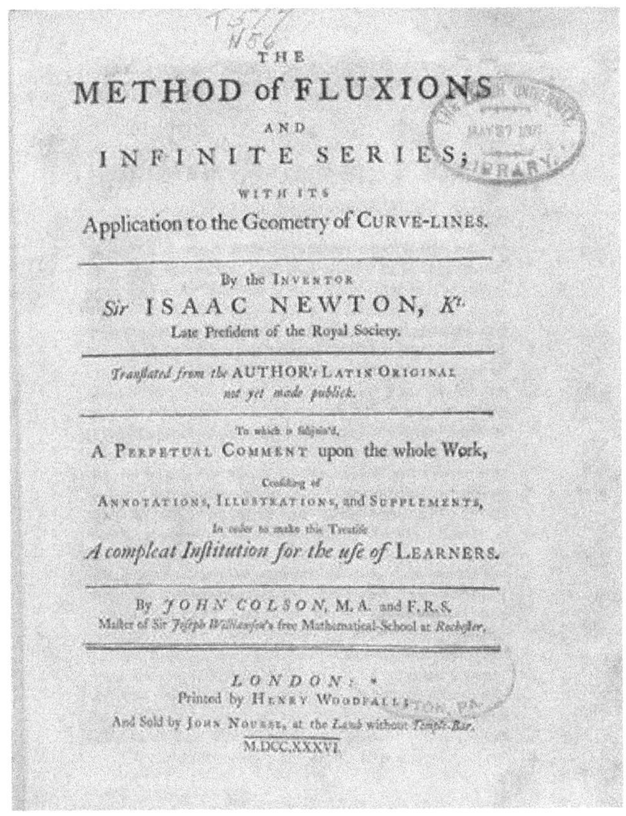

Figure 6.5   Newton's method of fluxions.

Gottfried Leibniz (Figure 6.6) was born three years after Issac Newton in 1646, in Leipzig, Germany. When he was only six years old, he lost his father, a professor of moral philosophy, and, like Isaac Newton, was raised by his mother. He was an extremely bright child who taught himself Latin and Greek by the time he was 12, having studied many of his father's books. At the Nicolai School in Leipzig, he was exposed to Aristotle's works on logic, which he felt were lacking, and therefore began to advance his ideas on how to improve Aristotle's system. At 13, he composed a three-hundred-verse Latin poem in one day, and at 14, he entered the University of Leipzig. This was not necessarily so unusual at that time, as there were other students just as young. His mathematics program was not very good, but he graduated with a bachelor's degree in just two years. In the summer of 1663, he reinforced his study of mathematics by studying with a professor at a school in Jena, Germany. His mother died the following year while Leibniz was back in Leipzig studying for his bachelor's degree in law. For some reason, perhaps his age, the University of Leipzig held back on granting him a doctorate in law, but his excellent scholarship enabled him to enroll immediately in the University of Altdorf, where he received a Doctor of Law degree in February 1667 at the age of 20.

Figure 6.6   Gottfried Wilhelm Leibniz.

While he worked in the field of law to improve the Roman civil law code for Mainz, Germany, his life took many turns. His abilities and his extraordinary range of interests produced projects in philosophy, mathematics, physics, theology, ethics, and even poetry in which he took great pride. He is considered by some to be the "last universal genius," because in subsequent years, with the advance of technology, it was difficult to excel in so many fields. In 1672, Leibniz went to Paris to make more scientific contacts and studied mathematics and physics under Christiaan Huygens (1629–1695). Huygens was a Dutch mathematician who patented the first pendulum clock and developed the foundations of mechanics. He studied the work of many recent mathematicians, including Pascal and Descartes, and his creative genius began to blossom. He was elected to the Royal Society of London as a Fellow in April 1673, and he began to study the geometry of infinitesimals that Newton had developed before his calculus was published. Leibniz developed a mechanical calculating machine, which he failed to finish, causing the Royal Society to consider him unfavorably. However, they were not aware that he was developing into a creative mathematical genius. He could be found sitting for several days in the same chair, thinking, as well as traveling throughout Europe all year, collaborating with many learned men. He corresponded with over 600 people from many different disciplines. He was a tireless worker, a true patriot, and an accomplished scientist who was one of the great contributors to Western civilization. In Paris, he began to create the fundamental concepts of his calculus. He spent much time trying to create a good notation, which he realized was fundamentally important to clarifying his mathematics. Notation is critical to clearly expressing and working with mathematical concepts. His notation for a derivative, developed by 1675, was so well expressed that it is still used today, 350 years later. Newton's notation, on the other hand, was written for his use and did not represent the concepts very clearly to many mathematicians. Leibniz's notation is as follows: Given y as a function of x, the derivative of y with respect to x is written in fraction form using d for derivative before the variables: $\frac{dy}{dx}$. The form of notation in mathematics can be very helpful to illustrate abstract ideas.

The other branch of calculus, Integral Calculus, also adopted Leibniz's notation to represent an *integral,* which is defined as the area under a curve. Here is an example that will help explain exactly what an integral is and how it is constructed. Let's look at the curve of a basic parabola, which has the simplest formula after a straight line. A parabola is the curve that an object traverses in a gravitational field. For example, a thrown baseball travels in a parabolic curve, and so does water traveling from a fountain. In these cases, the parabola opens down.

Consider now a parabola that opens up, whose equation is $y = x^2$, and the area under the curve down to the x-axis from $x = 1$ to $x = 2$. See Figure 6.7. To find this area, we draw a large number n of rectangles under this curve whose heights are the y values of the parabola and whose widths are a very small interval of x represented by dx.

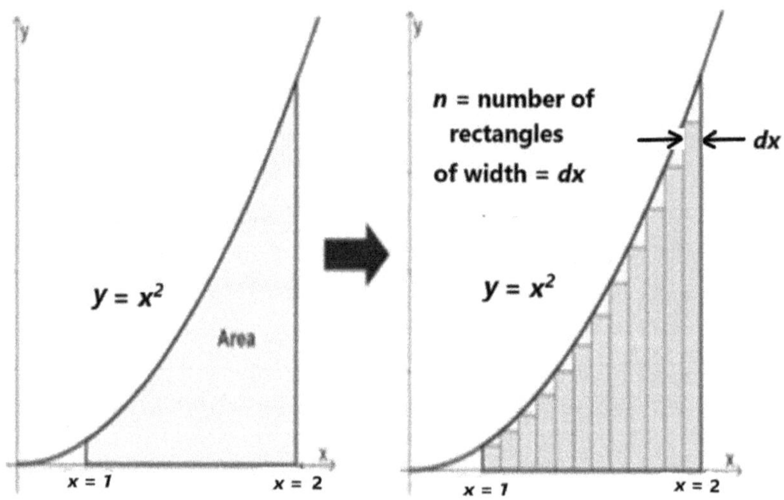

Figure 6.7   The integral of a function.

The area of each rectangle is then its length (value of y) times its width (dx) written (y)(dx). In this example, the value of y is equal to x^2. So, substituting x^2 for y, the area of each rectangle can be represented by (x^2)(dx) where x takes on the values from $x = 1$ to $x = 2$. We write this as $(x_i)^2\,dx$, where x_i represents some value of x between $x = 1$ and $x = 2$, and i takes on the values from 1 to n. If we add up all the areas of the rectangles, the total area will approximate the area under the curve. We express the total area by a summation sign, Σ (Greek capital sigma), and indicate that the value of i goes from $i = 1$ to $i = n$ as follows:

$$\textbf{\textit{Total area of the rectangles}} = \sum_{i=1}^{i=n} \left(x_i \right)^2 dx$$

Now if we increase the number n of rectangles, their widths dx will decrease and their total area will be a better approximation of the area under the curve. The area can be physically calculated for a large number of rectangles, just as Archimedes approximated the area of a circle, but here is where the power of calculus comes in. As the number of rectangles approaches infinity, and the width of each rectangle approaches zero, the total area of the rectangles approaches the actual area under the curve. The value that this area becomes infinitesimally close to, but does not exceed, can be determined exactly by the methods of calculus. We express this exact area using the integral sign created by Leibniz:

$$\textbf{\textit{Area under the parabola from }} x = \textbf{1} \textbf{\textit{ to }} x = \textbf{2} \textbf{\textit{ is expressed}}: \int_{1}^{2} x^2 dx$$

The integral sign is an elongated '*S*' representing an infinite summation. Using the methods of calculus, we can now easily determine the exact area under the parabola between $x = 1$ and $x = 2$. How to do this is miraculously revealed by *The Fundamental Theorem of Calculus,* discovered independently by the geniuses Newton and Leibniz. The Fundamental Theorem of Calculus is what ties together differential calculus and integral calculus. The process of evaluating an integral, which is not always easy, turns out to be the inverse process of finding a derivative. This was a major discovery relating two concepts which appear at first to be unrelated. A derivative can be found for almost any function, whereas an integral can only be evaluated for certain functions using the Fundamental Theorem. To evaluate this integral, we proceed as follows. The Fundamental Theorem requires that to evaluate an integral, we must figure out what function, when you apply the formula for its derivative, turns out to be the function in the integral. The formula for the derivative of any power n of x, written x^n, is just nx^{n-1}. The derivative is obtained by multiplying by the exponent and reducing the power by 1. For example, the derivative of x^2 is $2x^{2-1} = 2x$, the derivative of x^3 is $3x^{3-1} = 3x^2$, the derivative of x^4 is $4x^{4-1} = 4x^3$, and so on. The function in the integral is x^2. Notice that we have shown above that the derivative of x^3 is $3x^2$. We need to find a function whose derivative is not $3x^2$ but just x^2. We can do this by changing x^3 to the function $\frac{1}{3}x^3$. Now the derivative of $\frac{1}{3}x^3$ is $\frac{1}{3}$ times the derivative of x^3. This is calculated as follows: $\frac{1}{3}(3x^2) = x^2$. The function $\frac{1}{3}x^3$ is therefore the function whose derivative is x^2. This function, $\frac{1}{3}x^3$, is called the *antiderivative* of x^2 and is the function we need to evaluate the integral. The Fundamental Theorem tells us all we need to do now is substitute the two values of x, 2 and 1, into this antiderivative and find their difference. This is done as follows:

Area under the parabola from $x=1$ to $x=2$:

$$\frac{1}{3}(2)^3 - \frac{1}{3}(1)^3 = \frac{8}{3} - \frac{1}{3} = \frac{7}{3} \text{ or } 2\frac{1}{3} \text{ units}$$

This surprisingly simple calculation and rational result, which mathematicians struggled for centuries to find, is considered in mathematics to be an elegant solution to a challenging problem. Its elegance lies in its simplicity.

Leibniz produced his notation of the elongated S, $\int$, for the integral sign in 1675 and during the following year discovered some of the basic differentiation formulas. During this time, Newton wrote to Leibniz describing many of his results, but not his means of obtaining them. Leibniz must have realized at this time that he should publish a full account of his ideas. He replied to Newton, revealing some of his results; however, letters at this time could take as long as six months or more to be delivered. Newton expected to hear from Leibniz sooner, and wrote another letter to Leibniz on October 24, 1676, accusing him of stealing his methods. Leibniz did not receive this letter until June 1677. He replied, defending his ideas and giving details of some of his methods of differentiating functions. Leibniz did not conceive of a derivative as an infinite limit, which Newton correctly did, but mainly as a process to achieve a result. This prompted Newton to claim, with some justification, that Leibniz did not produce a solution to any previously unsolved problems. This prompted Leibniz to publish *Nova Methodus pro Maximis et Minimis, itemque Tangentibus* ("*A New Method for Maxima and Minima and Tangents*") in a journal in Leipzig (Figure 6.8). The publication contained various derivative formulas and his notation *dy/dx* for a derivative. A famous Swiss mathematician, Johann Bernoulli (1667–1748), criticized the paper for containing no proofs and called it an enigma lacking explanation. His work on integral calculus, *Acta Eruditorum*, appeared in the same journal in 1686 and contained the first appearance in print of the integral sign $\int$. Unfortunately, Newton's work did not appear in print until years later, in 1736, though there is little question that he had discovered the ideas of calculus before Leibniz. As with many significant discoveries, there comes a time when previous ideas provide the groundwork for major breakthroughs to emerge simultaneously by different people. This is exactly what happened with the discovery of calculus. Newton's failure to publish his results promptly led to significant consequences between England and Germany. In 1687, Newton published one of the greatest scientific books of all time, his *Philosophiae Naturalis Principia Mathematica* ("*Mathematical Principles of Natural Philosophy*") or *Principia* as it is commonly known today, in which he explained the motion of projectiles, pendulums, and bodies, both in free fall and under centripetal forces (Figure 6.9). He also formulated the law of attraction between heavenly bodies, including the planets

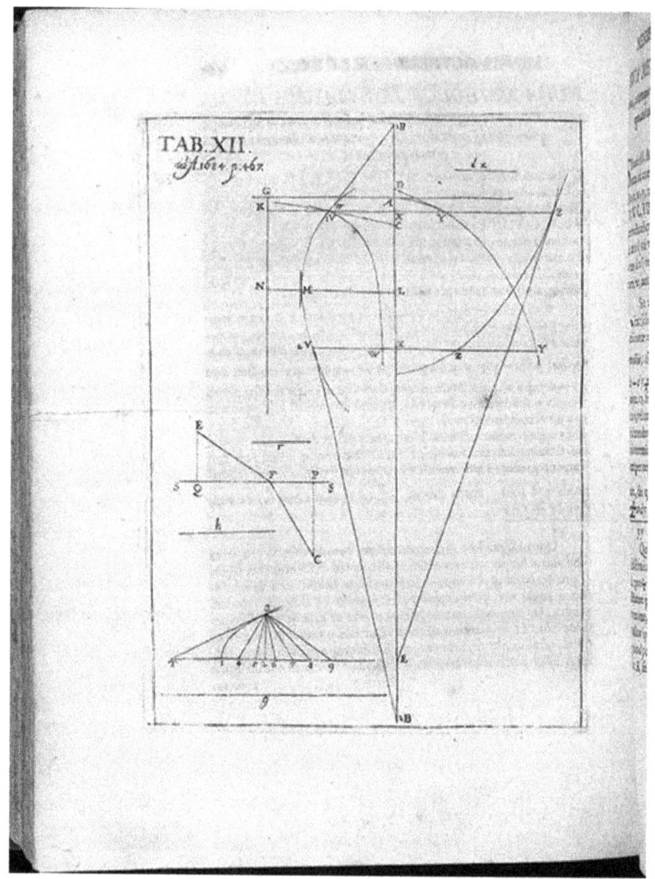

Figure 6.8   Diagrams opposite page 467 of Leibniz's calculus.[2]

toward the sun, and their elliptical motion, which provided the foundation for the Laws of Gravity. The French mathematician Guillaume L'Hôpital (1661–1704) wrote in 1696 that the ideas inherent in Newton's *Principia* were *"nearly all about this calculus."* Newton's calculus provides the foundation for his geometrical ideas in the *Principia*, but it lacked a clear explanation of the fluxional notation that he devised.

---

[2] Mathematical Association of America/Science Photo Library.

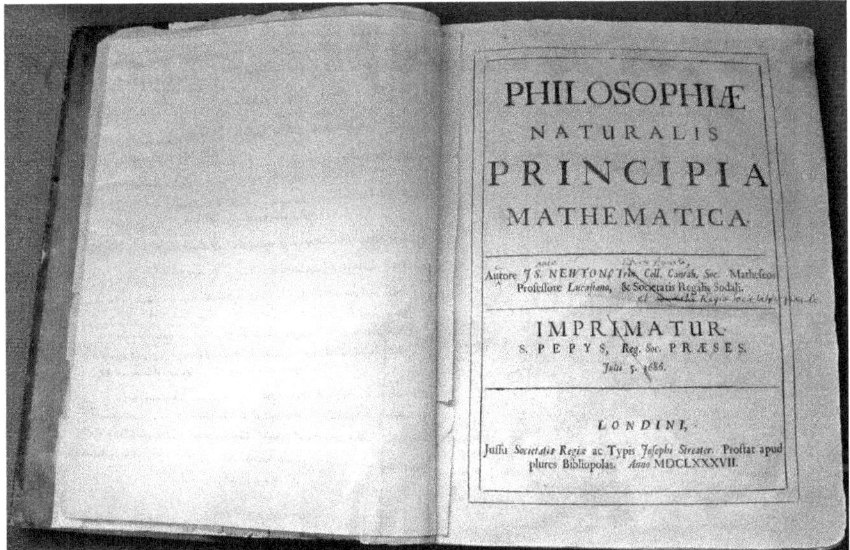

Figure 6.9   Newton's copy of *Principia* at Cambridge.

Priority of ideas was of great importance at this time, as it is today, to scientists and their discoveries. However, in the seventeenth and early eighteenth centuries, there were few scientific journals, and correspondence took a very long time to be delivered. There were few established procedures to determine the priority of scientific discoveries, such as the thorough peer review that exists today. To protect one's work, scientists used anagrams, hidden in sealed envelopes, private correspondence with trusted colleagues, and other means of concealing written material, such as secret codes that required a key to decipher. A letter could also be sent to a high official such as Henry Oldenburg (1619–1677), the secretary of the Royal Society of London, or Marin Mersenne (1588–1648), the founder of the French Academy of Sciences. The sender could prove he discovered the idea when the letter was sealed and stamped with the time, therefore absolving him of any plagiarism of a published article. Even when one took these steps, if similar material was later published by another person and put to important use, it could take priority over previously discovered material that was not utilized. It was also possible for a mathematician's work to be questioned as a mere improvement over facts that were already published and did not require great skill.

Scottish mathematician John Keill (1671–1721) took up Newton's cause with his blessing, and in 1708 accused Leibniz of plagiarizing Newton's work in the journal of the Royal Society. The Royal Society is a fellowship of many of the world's most eminent scientists and is the oldest scientific academy in continuous existence. Leibniz demanded a recantation, claiming he had not seen the calculus of fluxions until much later, but Keill pointed out that Leibniz had access to these ideas through the two letters sent to him by Newton. Leibniz insisted again that the Royal Society correct the harm done to him by Keill, so a committee of the Royal Society, of which Newton was the president at the time, was formed to formally investigate the dispute.

In any event, a bias favoring Newton tainted the whole affair from the outset. In response to a letter it had received from Leibniz, the Royal Society set up a committee to adjudicate on the priority dispute. That committee never asked Leibniz to give his version of the events. The report of the committee, *Commercium Epistolicum* ("*Epistolary Trade*") supported Keill's statement. However, because it did not consider Leibniz's version of the events it appeared very biased, and was written by Newton himself. It was published early in 1713, but Leibniz was not privy to it until the latter part of 1714. To help clear his name, Leibniz published a pamphlet called *Charta Volans* ("*Flying Paper*") defending his side and pointing out a mistake by Newton in his understanding of higher derivatives. This was found by the Swiss mathematician Bernoulli, who wrote that Newton could not have independently invented the calculus without seeing Leibniz's work. Keill then wrote to Newton:

> *I have read both pieces ... inserted in the* Journal Litéraire *and I think I never saw anything written with so much impudence, falsehood and slander as they are both. I am of the opinion that they must be immediately answered, and I am now drawing up an answer which I will finish as soon as I hear from you ...*

Keill continued the dispute and published a reply in Leibniz's *Charta Volans* about his claim, written to defend his national honor.

Leibniz called Keill an idiot and refused to continue the argument. Still, Newton was concerned and wrote to Leibniz directly, who replied with a thorough description of his discovery of the differential calculus. At this time, Newton's friendship with Keill began to wane as he grew tired of Keill pursuing the argument. Leibniz spent much of his last years trying to dispel the controversy and never agreed to accept that Newton had priority over the discovery of the calculus. However, most mathematicians in Britain during the 18th century sided with Newton. As expected in the German-speaking world, Leibniz was favored, though it cast a pall over his remaining life. These beliefs persisted until the 20th century, when historians began to recognize Leibniz's independent discovery and realized the differences between Newton's and Leibniz's versions of the calculus. Newton had created an infinitesimal calculus with clear definitions and algorithms. In contrast, Leibniz had created procedures and methods for calculating derivatives and integrals and devised a very useful and widely accepted notation. A monument In Oxford University credits both mathematicians for their independent contributions, Figure 6.10.

Figure 6.10 Newton and Leibniz in the Oxford University Museum of Natural History.

Leibniz's second major mathematical contribution, besides calculus, was his development of the binary system of arithmetic. The binary system, which contains just two numerals, 0 and 1, has important applications today in the world of computers and electronics. He published these results in 1701 in the paper *Essay d'une Nouvelle Science des Nombres* at the Paris Academy. Other important mathematical works were methods he devised to solve systems of linear equations, where he produced very effective notation and greatly simplified their solution. Very little of Leibniz's enormous amount of work was published in his lifetime. Scholars believe it could take at least fifty years to produce a comprehensive edition of his prodigious output, which could fill over a hundred volumes.

Both Newton and Leibniz possessed incredible talents; their lives and their work contributed to many fields other than mathematics. One of Leibniz's aims was to coordinate all the research of the learned societies, and he tried to collate all human knowledge. Though he made some progress, it was a task too difficult to achieve. He studied the research done on motion and published work supporting the German mathematician and astronomer Johannes Kepler's (1571–1630) treatise on planetary movement. Another major project he undertook in 1678–79 dealt with removing water from the mines in Hanover, Germany, using wind and waterpower to operate pumps. He designed many different types of windmills, pumps, and gears, but did not have successful results, which he blamed on the administrators and technicians. However, his observations and results were important scientifically, and he became one of the first scientists to study geology and correctly predicted that the Earth was molten when it was formed. He produced useful ideas in the field of dynamics and found issues with Descartes' notions of kinetic energy, potential energy, and momentum. However, his interest and ambitious work in many fields lacked the comprehensive contributions of his rival Newton, who managed to more clearly explain the concepts and express them on a more manageable level.

Leibniz devoted much of his life to trying to establish scientific societies in Berlin, Dresden, Vienna, and St. Petersburg. He was successful in promoting the Brandenburg Society of Sciences and became its president for life. Although the society was not very successful, it ultimately led to the creation of the Berlin Academy years later. He did provide some of the groundwork for the St. Petersburg Academy, but it was not established until after his death. He published significant work in philosophy arguing that the universe, though created by a good God, could not be perfect as it was subject to the laws of science and natural disasters. Thus, like in Voltaire's *Candide*, he claimed that this is the best of all possible worlds. One can safely say that Leibniz contacted most of the learned scholars in Europe and sought to foster mutual understanding across disciplines, unafraid to transcend boundaries between fields and offer novel perspectives on established dogma. He did not favor universities because he felt their environment did not favor the collaboration of specialties, which he felt was the essential catalyst for the advancement of knowledge.

Figure 6.11   Leibniz monument in Leipzig.

Leibniz made significant and worthwhile contributions to the fields of mathematics, physics, geology, logic, metaphysics, epistemology, philosophy of religion, jurisprudence, and history: A monument (Figure 6.11) in Leipzig honors Leibniz' genius and his contributions.

The French philosopher and art critic Denis Diderot (1713–1784), in his *Encyclopédie,* published in France between 1751 and 1772, had this to say about Leibniz:

Perhaps never has a man read as much, studied as much, meditated more, and written more than Leibniz. ... What he has composed on the world, God, nature, and the soul is of the most sublime eloquence. If his ideas had been expressed with the flair of Plato, the philosopher of Leipzig would cede nothing to the philosopher of Athens.

The irony of his life is that his contributions promoted more intellectual specialization and possibly added to scientific chauvinism, preventing the educated layman from easily grasping scientific developments. Sadly, Leibniz's funeral had only one mourner, his secretary and an eyewitness who wrote:

He was buried more like a robber than what he really was, the ornament of his century.

Sir Isaac Newton was one of the most brilliant scientists the world has ever known. He made contributions to optics, astronomy, fluid dynamics, alchemy, theology, and natural philosophy.

Besides his most famous work on gravitation, calculus, and other mathematical discoveries, he explained the tidal cycle, the Sun's effect on the motion of the moon, the abnormal orbits of comets, and the precession of the Earth's axis, which rotates around a small circle taking 26,000 years to complete one cycle. It took the learned community time to understand and accept that these movements could exist over such long distances, but his ideas eventually prevailed.

Newton constructed the first effective reflecting telescope (Figure 6.12), studied the spectrum through a prism, and developed an enlightened theory of color, leading to his well-received book *Opticks* in 1704. He was the first to make a calculated determination of the speed of sound. He proposed a basic mathematical model of a Newtonian fluid, which tends to flow uniformly, and whose viscosity remains somewhat constant. Water and air come close to being Newtonian fluids.

Figure 6.12   A replica of Newton's reflecting telescope in London's Science Museum.

Newton's life moved in various directions after his scientific achievements. He became the Lucasian Professor in Cambridge and was elected to represent and lead the resistance to the king's attempt to force Catholicism upon the University. He suffered from these encounters and other factors in his life, resulting in a couple of nervous breakdowns. These could have been caused by his religious beliefs, poisoning from his alchemy experiments, or frustration and criticism directed toward his research. He appears to have had bouts of depression during much of his later life. He became disenchanted with academia, gave up research, and sought a position in London. His friend, Lord Halifax, helped him to attain the appointment of Warden of the Royal Mint in 1696. He managed England's coinage very well, and upon the death of the present Master, Thomas Neale (1641–1699), he became Master of the Mint. He took the job seriously and sought to improve the currency and punish counterfeiters, holding the position for the remaining 30 years of his life. In 1701, he resigned his professorship at Cambridge and was elected President of the Royal Society in 1703. In recognition of his remarkable achievements, he was knighted by Queen Anne in 1705 (Figure 6.13). He was now

Figure 6.13   Monument for Newton in the Dutch Art Institute.

quite prominent and revered in London society and had close connections with individuals of power and authority. However, many of these last years were not without disputes over his research, and he was prone to display strong reactions at times, especially directed against Leibniz. It certainly is not easy living in the ordinary world when you are an exceptional genius, and we owe a great deal to this extraordinary man. Among his many quotes, two that especially show his humility toward his fellow human beings are worth remembering:

> *I do not know what I may appear to the world, but to myself I seem to have been only like a boy playing on the seashore and diverting myself in now and then finding a smoother pebble or a prettier shell than ordinary, whilst the great ocean of truth lay all undiscovered before me.*

If I have seen further than others, it is by standing upon the shoulders of giants.

On March 20, 1727, Isaac Newton died in his sleep and was the first scientist to be buried in Westminster Abbey among kings and queens.

Chapter 7

The Development of Group Theory

The power of calculus unleashed the development of mathematics, resulting in an expansion and abstraction of its concepts throughout the nineteenth century and beyond. Solutions were found to many previously unsolved problems, but at the same time, it spawned many more theoretical ideas to be wrestled with. Group theory is one of those expansive concepts that arose and provided a basis for analyzing different mathematical systems by finding a commonality in their structure. A group is an abstract concept that is defined as any set of elements or objects that includes only one closed operation between its members. A closed operation means that when one element is operated on another, it always produces another member of the group. The operation satisfies a few simple axioms that we illustrate by looking at the set of positive and negative integers: ... $-3, -2, -1, 0, 1, 2, 3,$ ... which form an infinite group under the operation of addition. Adding any two numbers *always* produces another member of the group. There is an *identity* element (zero) such that when added to any number equals the same number: $x + 0 = x$. For each number there also exists its negative, called the inverse, that when added to it results in the identity: $x - x = 0$. The only other axioms stipulated are commutativity, which means for any two elements a and b: $a + b = b + a$, and that of associativity, which means that for any three elements a, b, and c:

$$a + (b + c) = (a + b) + c.$$

The set of positive and negative integers under the operation of addition therefore satisfy the axioms for a group. These axioms that define a group are shown in Figure 7.1, where G represents the group and $G \times G \to$ G means it is closed under the operation. The circle **o** represents the operation, **e** represents the identity element, and g^{-1} represents the inverse of the element g.

$$\mathbf{o = operation}$$
$$\mathbf{o: G \times G \longrightarrow G}$$
$$\mathbf{a\ o\ (b\ o\ c) = (a\ o\ b)\ o\ c}$$
$$\mathbf{g\ o\ e = e\ o\ g = g}$$
$$\mathbf{g\ o\ g^{-1} = g^{-1}\ o\ g = e}$$

Figure 7.1   Group G axioms.

Many mathematical systems, other than numbers, in algebra and geometry, can be found to form a group. For example, three different coins in a row with the operation defined as a rearrangement, or permutation, of the three different coins without turning them over, form a group called a symmetry group. This is a finite group that includes six permutations where coins are rearranged and two permutations where they are not rearranged. The two permutations where they are not rearranged represent the identity and the inverse. The commutative and associative laws also apply. You can see that this simple illustration satisfies all the requirements of a group.

Group theory belongs to the field of abstract algebra whose early development is due to two young mathematicians: the Norwegian mathematician Niels Henrik Abel (1802–1829) and the French mathematician Évariste Galois (1811–1832). Tragically, each of their lives was cut short in their twenties, but their brilliance, like that of many other geniuses, shone when they were very young. Before we look more closely at the lives of these two extraordinary mathematicians, it is helpful to have a sense of the historical times in which they lived.

At the beginning of the nineteenth century, much work was being done in three areas of mathematics that sparked the development of group theory: geometry, number theory, and algebraic equations. These three areas were shown to exhibit the properties of groups. The first area, geometry, had evolved from its ancient roots and was now being studied more for its shape and form rather than its measurement, and for its properties in higher dimensions. An interesting example of this approach was a concept developed by the German mathematician August Möbius (1790–1868). In 1827, he began to focus on various geometric properties and attributes. His name is notably attached to an intriguing two-dimensional surface that he studied in 1865, a surface that possesses only one side and one edge. The surface is called a Möbius strip and can easily be formed by taking a belt and joining it together with one twist. If you now move your finger along one side, or one edge of the twisted belt, it will traverse the entire surface of the belt and return to where you started. This shows that the two sides are now one side, and the two edges are now one edge. The Möbius strip is therefore only a two-dimensional surface that exists in three dimensions. Such a property has inspired the creation of larger structures, which highlight the features of a Möbius strip. An example is shown in Figure 7.2.

The Möbius strip is an example of a geometric figure that belongs to the mathematical field of Topology, which emerged during the nineteenth century. Topology places geometric structures into groups that share certain minimal properties. For example, a baseball and an American football belong in the same topological group because one can be "flexibly shaped" into the other without distorting the geometry of the surface. Similarly, a ring and a doughnut are topologically equivalent. On the other hand, the Earth and a flat surface belong to

Figure 7.2 Möbius strip in Taichung city, Taiwan.

different topological groups because one cannot be shaped into the other without distorting the geometry. This can be seen in a flat map of the Earth, which distorts shapes near the poles. Cartographers always grappled with this difficulty while trying to produce the most useful flat map of the earth. Topological concepts find their way into almost all branches of mathematics, as well as into physics, biology, and robotics.

In the second active area of mathematics, number theory, in 1761, the Swiss mathematician Leonhard Euler (1707–1783) studied groups of numbers that have the same remainder when divided by a specific number. For example, consider the set of numbers with a common difference of 3:

$$1, 4, 7, 10, 13, 16, \dots$$

When any number in the set is divided by 3, the remainder is always 1. The number 3 is also a divisor of the difference of any two numbers in the set. For example, 3 divides $7 - 4 = 3$, $10 - 4 = 6$, $13 - 4 = 9$, etc. We say that any two numbers in this group, a and b, are equivalent to each other *modulo* 3. This is written $a \equiv b$ (mod 3). The sign $\equiv$ is stronger than an equal sign and means "equivalent to." The equal sign is used when we write an equation, such as $y + 2y = 5$, because it is true for *only* one value of $y = 2$. However, we use the equivalent sign when we write an identity such as $y + 2y \equiv 3y$, because it is true for *any* value of y. The expression $a \equiv b$ (mod 3), is true for all values of a and b. One side is considered identical to the other. This area of mathematics is called Modular arithmetic. It is sometimes called Clock arithmetic, because when you compare using time based on 12, to using military time based on 24, the following is true: $14 \equiv 2$ (mod 12) and $17 \equiv 5$ (mod 12). In each case, the hands of an analog clock are in the same position for both pairs of numbers, and the difference of each pair of numbers is 12.

It was the German mathematician Carl Friedrich Gauss (1778–1855; Figure 7.3) who demonstrated that Modular arithmetic contains the properties of a class of groups called Abelian groups, named after Niels Abel. Gauss lived a long productive life and was one of the most influential mathematicians of the nineteenth century. He has been called the "Prince of Mathematicians."

Figure 7.3. Johann Carl Friedrich Gauss.

Modular arithmetic is useful in many branches of mathematics, and also in chemistry, computer science, and cryptography. On a very practical level, International Standard Book Numbers. ISBNs, group numbers using either modulo 11 for a 10-digit ISBN or modulo 10 for a 13-digit ISBN as follows: If the ISBN is 10 digits, the sum of the digits must be divisible by 11, and if the ISBN is 13 digits, the sum of the digits must be divisible by 10. This allows a computer to check for the validity of the ISBN. To protect bank accounts, International Bank Account Numbers, IBANs, use up to 34 letters and numbers and employ modulo 97 arithmetic to detect an input error made by a user.

The brilliant mathematician Leonhard Euler (Figure 7.4) was not only the greatest mathematician of the eighteenth century, but one of the greatest mathematicians of all time. He was extremely prolific and made advances in many fields, even though he was blind for most of his productive life. To him we owe the notation π for pi, the Greek capital sigma Σ, for summation, the symbol $f(x)$ for "function of x", the Greek capital delta Δ, for difference, the letter e for the base of natural logarithms,[1] the letter i for the square root of -1, and several other notations used today.

Algebraic equations were the third active area of mathematics during the nineteenth century, which progressed after much work with them throughout the Middle Ages. Consider first the general quadratic equation, which is an equation of the second degree studied in algebra class:

$$ax^2 + bx + c = 0$$

where a, b, and c are the known coefficients. The following elementary formula (quadratic formula) can be used to solve any equation of this type when you substitute the values of a, b, and c:

$$x = \frac{-b \pm \sqrt{b^2 - 4ac}}{2a}$$

---

[1] The letter $e \approx 2.718...$ is a special irrational number like π and has been found to define a natural growth function. For example, when money is compounded daily, or actually every instant, \$1 at an interest rate of 10% will grow to \$2.718 $\approx e$ in 10 years.

Figure 7.4   Leonhard Euler.

This basic formula had been known for many years. However, the formulas to solve third-degree equations containing the term x^3 and fourth-degree equations containing the term x^4 had not been developed until much later during the Renaissance. These formulas were far more complex than the quadratic formula. As the power of x increases, the complexity of the solution increases exponentially. At the beginning of the nineteenth century, no one had been able to devise a formula that could solve every *general* fifth degree equation which contains six terms:

$$ax^5 + bx^4 + cx^3 + dx^2 + ex + f = 0$$

In 1770, the French mathematician Joseph-Louis Lagrange (1736–1813) discovered that the roots of algebraic equations, when arranged in different ways, possessed certain properties that were the foundations of group theory. However, he never related these properties to the concept of a group. The Italian mathematician Paolo Ruffini (1765–1822), studied equations of the fifth degree and, in 1799, was the first to prove the remarkable theorem that *no* formula exists to solve the general algebraic equation of the fifth degree. He drew upon LaGrange's work and created what are called groups of permutations. These are groups that contain all the possible arrangements of their elements, and he applied that to the terms of the fifth-degree equation. Ruffini's work was not generally accepted as there were some gaps in his proof, but his concept was correct.

Figure 7.5   Neils Henrik Abel.

The mathematician Niel Henrik Abel (Figure 7.5) was born prematurely in a rectory in Gjerstad in south-east Norway, where his father was a minister. The year 1802 was a difficult time for Norway, which was part of Denmark during the Napoleonic wars. Though Denmark tried to stay neutral, the British felt threatened, and after destroying most of the Danish fleet in 1801, they captured the entire remaining

Danish fleet in 1807. Denmark then joined the powers against Britain, resulting in Britian blockading Norway and an causing an economic crisis. Sweden then attacked Denmark in 1813, and a treaty was signed in 1814 giving control of Norway to Sweden. Young Abel endured this turmoil during his youth and was taught by his father, who was also active in politics, until he was 13 years old. He then attended the Cathedral school in Christiania (which is Oslo today), but the school was in a poor state, and Abel was not able to nurture his talent in mathematics and physics. However, in 1817, his fortune improved with the arrival of a new accomplished mathematics teacher, Bernt Holmboë (1795–1850), who tutored Abel in university-level mathematics. In the short span of a year, Abel absorbed the works of Euler, Newton, Lagrange, and other renowned mathematicians.

At age 16, Abel gave a rigorous proof of the binomial theorem which deals with the expansion of binomials to any power n: $(a + b)^n$. He proved it valid for all numbers, going beyond Euler's proof, which was only for rational numbers.

In 1820, fortune again turned against Abel at age 18 when his father died, leaving no money to continue his education. He was now burdened with the responsibility of supporting his mother and the rest of the family. Holmboë came to his aid and helped Abel to obtain a scholarship by raising money, which allowed him to remain in school and enter the University of Christiana one year later. He was at this time probably the most distinguished mathematician in Norway. While at the University, he met Christine Kemp, who soon afterwards became his fiancée. By the time he graduated in 1822, he had already worked on the solution of fifth-degree or quintic equations, which mathematicians had been working on for over 250 years. In 1821, he submitted a paper to the Danish mathematician Ferdinand Degen (1766–1825) in which he believed he had provided a solution to quintic equations. However, he soon discovered a mistake in his proof, but his ability garnered much support from Degen and other scientists, and he continued to do groundbreaking work in mathematics. His major contribution came in 1824 when he proved that radicals could not solve the general fifth-degree equation, which implied that there was no formula to solve the general fifth degree equation as there was for the second, third, and fourth degree equations. However, he did not advance the ideas of group theory related to the solution of equations. Abel knew about Ruffini's work and prefaced his proof with the following statement:

> *Geometers have occupied themselves a great deal with the general solution of algebraic equations and several among them have sought to prove the impossibility. But, if I am not mistaken, they have not succeeded up to the present.*

His proof clearly demonstrated that there is no one formula that exists to solve all possible quintic equations, like the quadratic formula, which can be used to solve all second-degree equations. This result of finding that there is *no* possible solution to a mathematical problem is just as critical as *finding* a solution to a mathematical problem. The proven fact that no general solution of the fifth-degree equation in terms of radicals can be found mathematically, does stir one's curiosity about the power and limit of mathematical reasoning, and the complexity of mathematical ideas. However, it is possible to solve any fifth-degree, or higher, equation by various other techniques, such as approximation, iteration, graphical methods, and other advanced means.

After publishing his result in a Danish journal in 1827, Abel traveled with colleagues to Denmark, Germany, Italy, and France to meet and discuss his work with other mathematicians. In France, he met with adversity and ran out of funds, his health suffered, he contracted tuberculosis, and he managed on only one meal a day. He returned to Norway and struggled to earn a living with his future wife while continuing to produce exceptional mathematics, even though his health continued to decline. His masterpiece showing the insolvability of the general quintic equation was submitted to the Paris Academy in a memoir, but was never acknowledged and possibly misplaced. As a biography of Abel describes:[2]

> *The paper was only two brief pages, but of all his many works perhaps the most poignant. He called it only 'A theorem': it had no*

---

[2] *Niels Henrik Abel: Mathematician Extraordinary*, O. Ore, New York, 1974.

introduction, contained no superfluous remarks, no applications. It was a monument resplendent in its simple lines — the main theorem from his Paris memoir, formulated in few words.

Figure 7.6   Memorial plaque to Abel unveiled in 2014 in Berlin.

On April 8, 1829 a friend succeeded in obtaining an appointment for him in Berlin, but it was too late. Abel had died of tuberculosis on April 6, at the age of only 24, enduring much agony during his final days. This was indeed a tragic loss of a brilliant mathematician. His Paris memoir was found in 1830 and printed in 1841, but curiously disappeared again and turned up more than a century later in Florence in 1952. In 1830, the Paris Academy awarded Niels Abel and the German mathematician Carl Jacobi (1804–1851) the Grand Prix for their joint outstanding work on elliptic functions. Though Abel's work was Nobel prize material, the Swedish chemist Alfred Nobel (1833–1896) did not provide for a prize in mathematics for several reasons. As an industrialist and the inventor of dynamite, Nobel's will specified that the prizes should go to those who "conferred the greatest benefit on mankind," and perhaps he considered mathematics not practical enough. Also at the same time, a prize for mathematics was awarded by the King of Sweden, which may have influenced Nobel's decision. In 2001, the King of Norway established the Abel Prize (Figure 7.7), a prestigious award akin to the Nobel Prize, in honor of Niels Henrik Abel. It was first awarded in 2003 and is issued with a monetary award of 7.5 million Norwegian krona. The Fields Medal, awarded to mathematicians younger than age 40 by the International Mathematical Union, is also equivalent to the Nobel Prize.

Figure 7.7   Abel Prize.

Equally tragic was the life of the other brilliant mathematician Évariste Galois (Figure 7.8), who deserves the main credit for the development of early Group theory. He was born in 1811 when Napoleon was at the pinnacle of his power, nine years after Niels Abel and only 22 years after the storming of the Bastille in 1789. France experienced many years of political unrest during his young years. Both his parents were gifted and well-educated, but they did not display much ability in mathematics. His mother tutored him in Greek, Latin, and religious doctrine until he was 12 years old, while his father served as the Mayor of Bourg-la-Reine in Paris, France. Napoleon abdicated his rule in 1814, after failing in his Russian campaign, and Louis XVIII was installed as the King. Napoleon tried again in 1815 to regain power but was defeated at Waterloo in that year, and Louis XVIII was reinstated on the throne. Louis XVIII died nine years later and was succeeded by King Charles X.

At age 12, during these political conflicts, Galois enrolled at the Lycée Louis-le-Grand and did well in school, receiving several prizes. His rhetoric, however, was weak, and he had to repeat one of his last school years. Soon after that, his passion for mathematics began to emerge, and his director of studies wrote:

It is the passion for mathematics which dominates him, I think it would be best for him if his parents would allow him to study nothing but this, he is wasting his time here and does nothing but torment his teachers and overwhelm himself with punishments.

Figure 7.8   Évariste Galois.

His brilliance was recognized, but he was considered bizarre and removed, while his methodology was vague and found to be lacking in his work. At age 15, he had mastered advanced geometry and algebra written for professional mathematicians, but was uninspired by his daily classwork. At age 17, Galois failed his first examination for the prestigious University of Paris, the École Polytechnique. Besides the academic attraction of the University, Galois was also keen on the passionate political environment of the school due to his parents being active Republicans. When he returned to Louis-le-Grand, he enrolled in the mathematics class of Louis Richard (1795–1849), who noted that Évariste worked only in the highest realms of mathematics.

In 1829, he submitted several mathematical articles to journals and published his first mathematics paper in the *Annales de Mathématiques*. This year also brought great misfortune as his father, the mayor, became the political victim of damaging epigrams promoted by the priest of Bourg-la-Reine directed against him and his family. His father meant no harm, and the scandal that developed caused him great suffering and led him to take his own life. The loss of his father had a profound influence on Galois's remaining years. He applied again to the École Polytechnique, but given his circumstances, he struggled to effectively communicate his mathematical ideas, and he failed again. Galois then entered the École Normale in Paris, completing his degree within the year. Reports by his literature and mathematics examiners demonstrate his contrasting abilities:

Literature Examiner:

This is the only student who has answered me poorly, he knows absolutely nothing. I was told that this student has an extraordinary capacity for mathematics. This astonishes me greatly, for, after his examination, I believed him to have but little intelligence.

Mathematics Examiner:

This student is sometimes obscure in expressing his ideas, but he is intelligent and shows a remarkable spirit of research.

Such were the internal struggles that plagued the genius Galois. He continued his mathematical research on the theory of equations, but he discovered that some of his work had been covered in a posthumous article by Abel who died when Galois was age 18. This caused him to revise his approach, and he submitted a new article to the Paris Academy in February 1830 for the Grand Prix in mathematics. His work, unfortunately, was lost and never recovered. The Grand Prix was awarded to Abel that year, but despite Galois' lost work, he published three papers, one of which contained the beginnings of group theory.

In 1830, political revolution erupted again in France (Figure 7.9) against the conservative policies of King Charles X, who ultimately was forced to flee the country. Galois, along with all the students, was locked in the École Normale by the director to prevent them from taking part in the revolution. Galois tried to scale the wall to join the rioting in the streets of Paris but was unsuccessful. The director of the school was taunted by the students and retaliated by writing

Figure 7.9 July revolution of 1830[3].

---

[3] Copyright: The Collector.com.

articles in the newspaper attacking their actions. Galois responded by writing in the *Gazette des Écoles* and assailed the director for locking the students in the school. This resulted in Galois's dismissal from the École, whose life was further complicated by the death of his mentor, the French mathematician and physicist Joseph Fourier (1768–1830). Fourier is generally credited with the discovery of the atmospheric greenhouse effect. Galois' situation was described in a letter written by the French mathematician Sophie Germain (1776–1831):

> *... the death of Fourier has been too much for this student Galois who, in spite of his impertinence, showed signs of a clever disposition. All this has done so much that he has been expelled from the École Normale. He is without money, and they say he will go completely mad. I fear this is true.*

Galois' strong liberal political leanings led him to join the Republican branch of the French militia called the *Artillerie de la Garde Nationale* ("Artillery of the National Guard"). The Guard attempted to reverse the establishment of the Louis Philippe monarchy. However, the king disbanded the Artillery of the National Guard at the end of 1830, as it was a threat to the throne, and arrested 19 officers. The officers were acquitted, and the Republicans celebrated with a dinner. At the dinner, Galois raised his glass and a dagger and appeared to make accusations against the King. He was arrested after dinner but was acquitted by a jury a month later, possibly because his actions were thought to be the result of an impertinent youth.

Galois tried to return to mathematics during this time and published two articles in December 1830 and January 1831, which were the last ones in print during his life. He was the first to fully comprehend that the algebraic solution of an equation was tied to the structure of a group consisting of permutations associated with the roots of an equation.

More political problems arose on the following Bastille Day (July 14, 1831) when Galois was arrested for carrying several pistols, a loaded rifle, and a dagger. He was at the head of a protest wearing the uniform of the disbanded artillery. His stay in prison was wrought with depression, and he resorted to alcohol for the first time. While drunk, he confided to one of the inmates that

> *I will die in a duel because of some coquette de bas étage (low-class coquette) who will invite me to avenge her honor which somebody has compromised. ... I lack someone whom I can love.... I've lost my father and no one has ever replaced him.*

While in this state, Galois attempted suicide but was prevented by his fellow inmates. Fearing his future, he worked on all his mathematical manuscripts while in prison and was released almost a year later on April 29, 1832. Though it is not clear at this time what caused Galois to be faced with a duel, it was to take place one month later, on May 30. We know that five days before the duel, he wrote a letter to a friend mentioning an unhappy love affair. Fragments of letters written to Galois by Stéphanie-Félicie Poterin du Motel, a woman he knew at the hostel where he was staying, suggest that she confided some of her troubles to him. Other letters that Galois wrote to friends the night before the duel seem to support this conjecture and may have caused him to provoke the duel to avenge her situation. It is also not clear who his opponent was in the duel. He may have been one of Galois's Republican colleagues or the fiancé of Poterin du Motel. The reasons for the duel are lost to history, but Galois was so sure of his impending demise that he spent the entire night before the duel writing letters to his Republican friends. To his good friend Auguste Chevalier, he composed *La Lettre Testament*, his last mathematical testament consisting of three manuscripts of mathematics that contained the essence of his life's work in Group Theory. The opening words of his testament are shown in Figure 7.10. The German mathematician and physicist Hermann Weyl (1885–1955) had this to say about Galois's testament:

> *This testament, if judged by the novelty and profundity of ideas it contains, is perhaps the most substantial piece of writing in the whole literature of mankind.*

> ### The Last Mathematical Testament of Galois
>
> *"My dear friend, I have analyzed several new ideas. One concerning the theory of equations; the other, integral functions. In the theory of equations, I have studied as to in which cases the equations are solvable by radicals, which has provided me with an opportunity to go into this theory in depth and describe all possible transformations on an equation, even when it is not solvable by radicals. All this can be put in three papers. The first one is written, and, in spite of what Poisson has said, I stand by it, with the corrections that I have indicated. The second contains rather interesting applications from the theory of equations. ...*

Figure 7.10   Introduction to Galois's *La Lettre Testament.*

Soon after dawn on May 30, 1832, Galois was shot in the abdomen and was tragically left to die by his opponent and by both seconds, whose role was to try to reconcile the dispute without violence. Galois was later found by a passing farmer and taken to the Hôpital Cochin. He rejected the solace of a priest and succumbed the following morning, most likely to peritonitis. Unfortunately, he might have been saved if he had been cared for sooner. His last words to his younger brother Alfred were:

Ne pleure pas, Alfred! J'ai besoin de tout mon courage pour mourir à vingt ans ! (Don't weep, Alfred! I need all my courage to die at twenty!)

An uprising occurred a week after his death, protesting his unnecessary loss as a political activist. Figure 7.11 shows the Galois memorial located in the cemetery of Bourg-la-Reine, which was his native town in the southern suburbs of Paris. Galois was buried in a common grave in the Montparnasse Cemetery, but unfortunately, the exact location is not known.

Évariste Galois created one of the significant foundations of modern algebra that is embraced today. In September 1832, his faithful friend, Auguste Chevalier, published his testamentary letter in the *Revue Encyclopédique,* which was Galois's request. Chevalier believed that if the École Polytechnique had accepted Galois's work, he might have lived and not have been drawn to the political activism of the

Figure 7.11   Galois memorial.

Republicans, but would have been accepted as an accomplished mathematician. Copies of two of the most important unpublished works of Galois were made by Chevalier and, in 1843, given to the mathematician Joseph Liouville (1809–1882). In 1846, Liouville published the Oeuvres *Mathématiques d'Évariste Galois* in his *Journal de Mathématiques Pures et Appliquées*, but did not fully grasp the fundamental importance of the group concept. It is not clear why ten years passed before Chevalier shared Galois's work, permitting it to be published for the first time. Galois's political activism and the turmoil in France could have contributed to the lack of widespread acceptance and publication of his work.

Évariste Galois's groundbreaking contribution was his unique proof of the insolvability, using radicals, of not just the general equation of the fifth degree but *all* higher degree equations. This result depended upon what is now called Galois Theory, which has far-reaching applications in most branches of mathematics.

And so, we have the stories of two brilliant men, Niels Henrik Abel and Évariste Galois, who made significant mathematical contributions while still so young, but then had their lives needlessly taken by the times. These gifted mathematicians, in a more modern world, could have lived to give us much more.

Chapter 8

The Breakthrough of Non-Euclidean Geometry

One of the great revelations of the nineteenth century was the emergence of geometries that did not obey the laws of the ancient Greek geometer Euclid. Euclidean, or plane, geometry is the geometry of a flat surface, and was one of the great contributions of ancient Greece. It was the mathematical bible for over two thousand years and assumed almost a holy acceptance by the Church and many scholars and scientists. For most people, their small world appeared to be a flat surface, and Euclidean geometry served them well. However, as discussed in Chapter 2, the ancient Greeks did observe that the world was spherical and calculated the circumference of the Earth to an amazing accuracy of 98%. The geometry of the Earth, which is very different from plane geometry, is a type of non-Euclidean geometry called spherical geometry. It did not receive mainstream acceptance for many years as a legitimate type of geometry due to the church's influence and the dominance of Euclidean geometry. The great German philosopher Immanuel Kant (1724–1804; Figure 8.1) believed that Euclidean geometry transcended empirical knowledge and is inherent in the way we perceive space. He stated:

Euclidean Geometry is the inevitable necessity of thought and that their laws were immutable.

Figure 8.1   Immanuel Kant.

The vast number of ideas and theorems of Euclidean geometry are based on just five postulates or axioms, which are assumed true:

1. *To draw a straight line from any point to any other.*
2. *To produce a finite straight line continuously in a straight line.*
3. *To describe a circle with any center and distance.*
4. *That all right angles are equal to each other.*
5. *That, if a straight line falling on two straight lines makes the interior angles on the same side less than two right angles, if produced indefinitely, they intersect on the side on which the angles are less than the two right angles.*

The last or fifth postulate, which is illustrated in Figure 8.2, is essential to Euclidean geometry but requires some effort to fully realize its importance. In Figure 8.2, angle A is a right angle, which is equal to 90°, but angle B is an acute angle, which is less than 90°. Therefore, the sum of angle A plus angle B is less than two right angles and, as the fifth postulate states, the two lines will intersect on the side of angles A and B when extended indefinitely. For many years, mathematicians tried to eliminate the fifth postulate and reduce the number of postulates to only four. They tried to prove that the fifth postulate can be deduced from the first four, but all the "proofs" were eventually shown to contain errors. These false proofs further supported the necessity of the fifth postulate. This was indeed a strong testament to the brilliance of the ancient Greeks, who realized the need for this important axiom.

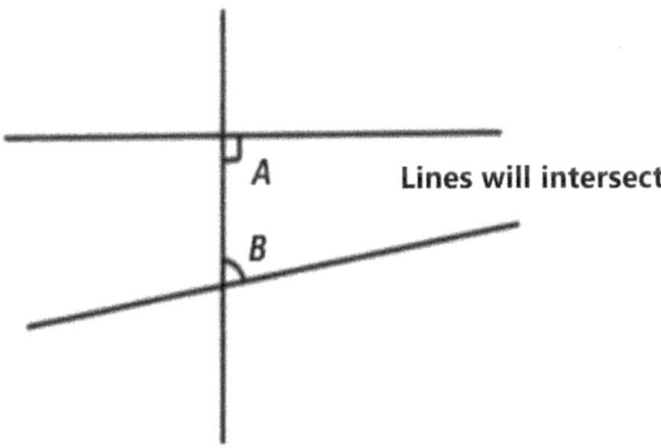

Figure 8.2   Fifth Euclidean postulate, angles $A + B < 180°$.

We can get a better understanding of the meaning of the fifth postulate from the following axiom developed in 1795 by the Scottish mathematician John Playfair (1748–1819). It is equivalent to the fifth postulate:

5. ***Given a straight line and a point not on the line, it is possible to draw one and only one line through the given point parallel to the given line.***

Playfair's axiom, as it is known, helps to provide a clear picture of what distinguishes Euclidean geometry from non-Euclidean geometry. If we accept the first four postulates and change Playfair's axiom in two basic ways, we arrive at the two fundamentally important geometries that are non-Euclidean: hyperbolic and elliptic. If we assume that more than one line can be drawn parallel to the given line, it creates hyperbolic geometry. Conversely, if we assume that no line can be drawn parallel to the given line, it creates elliptical geometry. The geometry of the earth, spherical geometry discussed above, is a special case of elliptical geometry. These two non-Euclidean geometries are discussed in more detail below.

As previously mentioned, many mathematicians tried to deduce the fifth postulate from the other four, but without success, because it is essential to the basis of Euclidean geometry. The non-Euclidean geometries did not develop and become widely accepted until more than 2300 years after Euclid. As with other epiphanies in science, art, and mathematics, the state of knowledge and the nature of the times often create an environment for discoveries or developments to occur simultaneously and independently from more than one source. Calculus developed in this way as shown in Chapter 6, and non-Euclidean geometry also evolved through the efforts of several mathematicians working independently at about the same time in the 19th century.

Hyperbolic Geometry

One of the greatest and most prolific mathematicians of all time, the German genius Carl Friedrich Gauss (1777–1855; Figure 8.3) entertained thoughts in his early years of the existence of non-Euclidean geometry. Gauss was an extremely productive mathematician and contributed to many disciplines, including number theory, geometry of surfaces, analysis, astronomy, magnetism, geodesy, and optics. His work was very influential in many of these areas, and his name is used to describe a process called *degaussing*, which is the destruction of data on a storage device done by altering or removing the magnetic field. At the age of seven, Gauss astounded his teachers by instantly adding all the numbers from 1 to 100 much faster than his classmates. He keenly observed the following: When one adds the first number, 1, and the last number, 100, the sum is 101. When one adds the second number, 2, and the next to the last number, 99, the sum is also 101, and when one adds the third number, 3, and the third from the last number, 98, the sum is again 101. So when one pairs all numbers from 1 to 100 that add to 101, you have 50 pairs of numbers, and therefore the

Figure 8.3   Statue of Carl Friedrich Gauss in Brunswick.

sum of all the numbers is simply $50 \times 101 = 5050$. This is an example of an arithmetic progression where the difference of successive terms is always the same. For example, the following series 2, 4, 6, 8, 10, … is an arithmetic progression where the common difference is 2. The series 10, 20, 30, 40, 50, … is also an arithmetic progression where the common difference is 10. Gauss's calculation of the sum of the numbers from 1 to 100 applies the basic formula for the sum S of any arithmetic progression when you know the first term a, the last term l, and the number of terms n:

$$S = \frac{n}{2}(a+l)$$

Gauss grew up in Brunswick (now the German state of Lower Saxony), and his family was not of a very high social status. His father, Gebhard, was a common laborer who worked at several trades, bricklayer, butcher, and gardener, while his mother was practically illiterate. Gebhard was proficient in writing and calculating and worked for a time as a treasurer for a financial fund. Gauss respected his father but felt he was dominating and somewhat callous at times. His elementary teachers, realizing he was a child prodigy, introduced him to the Duke of Brunswick, who helped him financially to enter the Collegium Carolinum. This institution became the Technical University of Braunschweig. He was there for three years, after which the Duke enabled Gauss to enter the University of Gottingen at age 18, where he excelled in mathematics, the sciences, and classical languages. Abraham Kästner (1719–1800), the German mathematician who compiled encyclopedias and wrote textbooks, was one of Gauss's professors. Gauss described him as

A leading mathematician among poets and a leading poet among mathematicians.

He called him a poet because of his witty epigrams and brief satirical statements, which were often in the form of poems. At the University, Gauss became friendly with the Hungarian student Farkas Bolyai (1775–1856; Figure 8.4), with whom he developed a lifelong friendship. In their correspondence, they explored the existence of a non-Euclidean geometry and the fact that proofs that tried to deduce the fifth axiom of Euclidean geometry from the other four lacked accuracy. Gauss also corresponded with the German mathematician and astronomer Heinrich Christian Schumacher (1780–1850), in which he admitted that his reputation would suffer if he made known his belief that there existed non-Euclidean geometry. Farkas Bolyai went on to teach mathematics, chemistry, and physics at Calvinist College in Marosvásárhely, Hungary (now Târgu-Mureş, Romania). His son János Bolyai (1802–1860) proved to be an extremely bright and talented child. At the age of 4, he could identify many geometrical figures, recognize the most common constellations, and was able to understand the trigonometric sine function. János practically taught himself to read a year later at the age of 5 and mastered the violin at age 7. His father keenly wanted him to become a mathematician and kept him from entering school until the age of 9 so he could teach him mathematics. At age 13, János became proficient in calculus and other branches of analysis. He skipped several years in Calvinist College, and when he was age 14, his father asked Gauss if he would tutor his son so that he could become extremely proficient in mathematics. Though it could have been an amazing opportunity for János, Gauss did not accept the request. After he graduated from Calvinist College at age 15, Farkas could not afford to send his son to a top-quality university. János remained at the College for one more year so that he could demonstrate high enough achievement to enter the best option, the Imperial and Royal Military Academy in Vienna, at the highest level. The Academy provided a quality education in mathematics, and János was commissioned as a sub-lieutenant four years later, having completed the seven-year course in just four years. He excelled in sports and became an accomplished violin player performing in Vienna. However, his strong will made it difficult for him to accept strict military discipline. He did not smoke or drink and was reputed to be the best dancer and swordsman in the Austro-Hungarian Imperial army, becoming a captain at age 24 and serving a total of 11 years. In addition to these skills, he was proficient in nine foreign languages, including German, Latin, French, Italian, Romanian, Tibetan, and Chinese.

Figure 8.4   Farkas Bolyai (left) and János Bolyai (right).

From the time he was age 18, throughout his years in military service, János could not stop thinking about Euclid's parallel postulate, and by altering it, what effect it would have on the geometry. His father, who had studied the same ideas for many years, wrote to him in 1820:

You must not attempt this approach to parallels. I know this way to the very end. I have traversed this bottomless night, which extinguished all light and joy in my life. I entreat you, leave the science of parallels alone... Learn from my example.

However, János was obsessed with pursuing the parallel postulate and its significance. He eventually discovered that negating the postulate could produce other consistent geometries. During the three years from the age of 18 to 21, he composed a treatise on parallel lines that he later called *absolute geometry*. He wrote to his father in 1823:

I have discovered such wonderful things that I was amazed... out of nothing I have created a strange new universe.

In 1825, János traveled to see his father and explain his discoveries to him; however, he was disappointed when Farkas did not show much enthusiasm for his ideas. It may be that his father did not want to encourage his son to publish his work. However, six years later, Farkas realized the full significance of his son's accomplishments and urged him to compose his work and publish it as an Appendix to the *Tentamen*, which was a journal written for young people to learn about pure mathematics. János was so pleased that he later wrote:

Had my father not happened to urge or even force me to immediately put things to paper, possibly the contents of the Appendix would never have seen the light of day.

In 1831, the Appendix was published and contained a brilliant exposition of two different geometries. Euclidean geometry and *absolute geometry*. Absolute geometry assumes just the first four postulates of Euclidean geometry and not the fifth postulate. It has limited application and geometric proofs, but provides the basis from which the three fundamental geometries, Euclidean, elliptical, and hyperbolic, are built upon. János used his absolute geometry to develop hyperbolic geometry, which assumes as a fifth postulate:

Given a line, more than one parallel line can be drawn through a point not on the given line.

Euclidean geometry is the geometry of a plane, or flat surface, having *zero curvature*. Non-Euclidean geometry is the geometry of a curved surface. There are two basic types of curved surfaces, convex and concave. Elliptical geometry is the geometry of a *convex surface that has positive curvature*. Hyperbolic geometry is the geometry of a *concave surface that has negative curvature*. Figure 8.5 shows a spherical surface that has constant positive curvature compared to a hyperbolic surface that has constant negative curvature. It is not possible to draw a straight Euclidean line on either of these two surfaces. However, the vertical lines shown on these surfaces are analogous to the straight lines of Euclidean geometry. They possess the important characteristic that a straight line in Euclidean geometry possesses, which is the unique property of being the *shortest distance between any two points* on the surface.

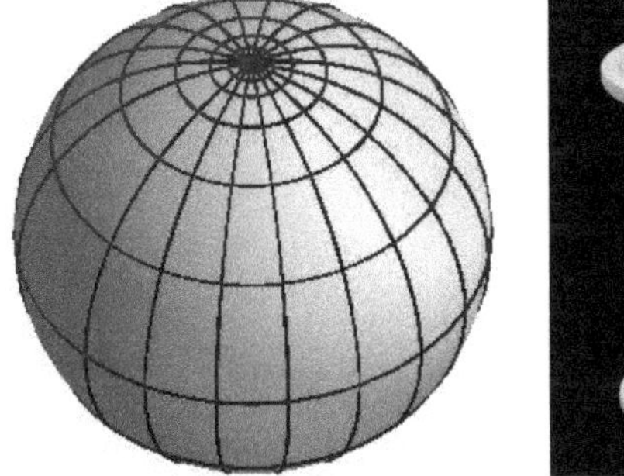

Figure 8.5   (a) Spherical surface of constant positive curvature; (b) hyperbolic surface of constant negative curvature.

A line that possesses this property on a curved surface is called a *geodesic* and serves the same function as a straight line on a plane surface. Consider the sphere in Figure 8.5, which is formed by rotating a circle 360 degrees. The "straight lines" or geodesics that are shown on the spherical surface are the great circles that pass through the two poles of the sphere. Notice that many geodesics pass through the two poles, which means there are many shortest paths between these two points, unlike Euclidean geometry, where there is only one shortest path between any two points. It turns out that on a convex surface, given a line (geodesic) and a point not on the line, it is not possible to draw a line through the point that is parallel to the given line (will not intersect). All geodesics (great circles) on the sphere intersect each other.

Hyperbolic Geometry

Now consider the hyperbolic surface in Figure 8.5. The vertical lines are geodesics, which are all hyperbolas. A hyperbola is the curve shown in Figure 8.6 on a plane surface. The hyperbolic surface is formed by rotating this hyperbola in a circle about its axis. In Figure 8.5, the hyperbola can be seen by the shape of the sides of the hyperbolic surface.

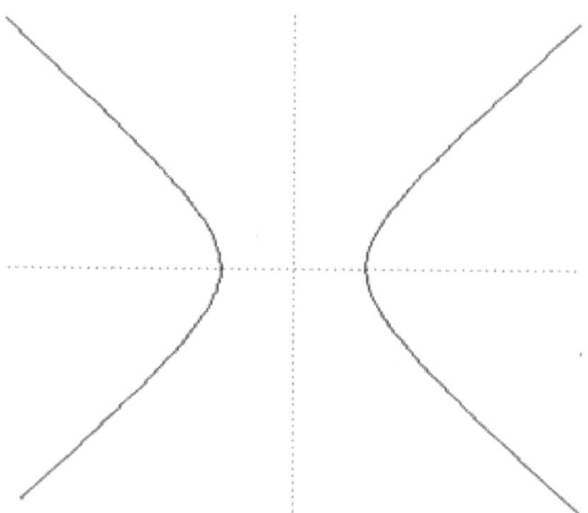

Figure 8.6   Hyperbola on a plane surface.

The "lines" or geodesics on the hyperbolic surface are the shortest distance between any two points on the surface. Notice how the lines diverge from each other in both directions. On a hyperbolic surface, given a line and a point not on the line, it is possible to draw many lines through the point that will not intersect the given line. This is because the curvature is such that any line drawn through the point slowly diverges from the given line. One consequence of this property is that if two lines make right angles with a third line, no matter how far they are extended, they will not intersect, and they will gradually move away from each other. We measure an angle between two intersecting curved lines by the angle formed by the curve's tangents to the lines at the point of intersection. The lines that intersect in Figure 8.5 all make right angles with each other.

Another interesting example of a hyperbolic surface is the one shaped like a saddle shown in Figure 8.7. Hyperbolic lines that diverge are shown on that surface. Hyperbolic geometry differs from Euclidean geometry in the following ways. In plane geometry, the circumference of a circle is equal to $2\pi r$, where r is the radius, and the sum of the angles of a triangle is equal to 180°. In hyperbolic geometry, the circumference of a circle is *greater* than $2\pi r$, and the sum of the angles of a triangle *is less than* 180°.

János Bolyai's father, Farkas, sent a reprint of the Appendix that János published in 1831 to Gauss, who, upon reading it, wrote to a colleague stating:

I regard this young geometer Bolyai as a genius of the first order.

However, to his old friend Farkas, Gauss wrote:

To praise it would amount to praising myself. For the entire content of the work... coincides almost exactly with my own meditations which have occupied my mind for the past thirty or thirty-five years.

The fact remains that Gauss was reluctant to compromise his reputation and publish these ideas, even though in 1824 he wrote that he had discovered a consistent non-Euclidean geometry.

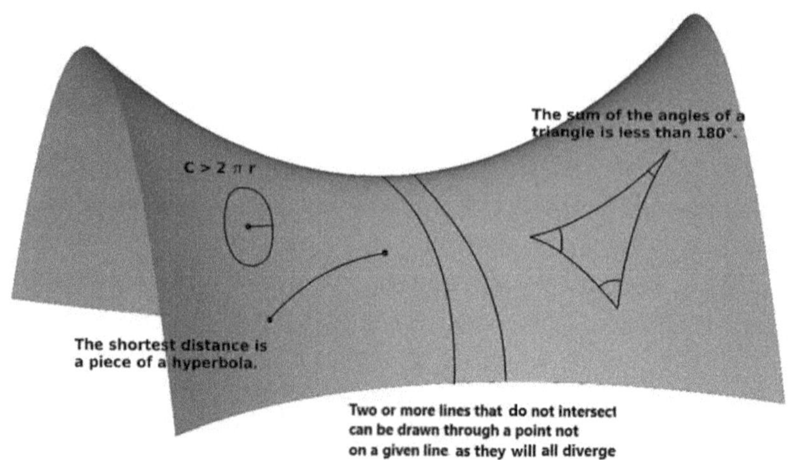

Figure 8.7   Hyperbolic saddle surface.

When János saw Gauss's letter, he could not help suspecting that his father had informed Gauss of his discoveries, causing ill feelings toward his father and strong anger at Gauss's attitude. There is little doubt that Gauss independently discovered non-Euclidean geometry close to the time that Bolyai had discovered it. In 1824, Gauss wrote a letter to the German mathematician Franz Adolph Taurinus (1794–1874) revealing his work with non-Euclidean geometry and many of the ideas associated with it, including the fact that the sum of the angles of a triangle is less than 180°. Unfortunately, this experience caused János to become greatly upset and have difficulty working. His health suffered, and he developed a fever at times, which made it strenuous for him to carry out his military duties. On June 16, 1833, he requested his pension and retired from military duty, going to live with his father for a while. He struggled the rest of his life, and though he only published a few pages of the Appendix, he left unpublished more than 20,000 pages of mathematical work when he died of pneumonia at the age of 57. However, János Bolyai deserves credit for his independent discovery and the first publication of non-Euclidean or hyperbolic geometry.

Nikolai Ivanovich Lobachevsky

Meanwhile, in a different part of the world, while Bolyai was developing his lifelong work on hyperbolic geometry, another mathematical prodigy arose who also envisioned a world of non-Euclidean geometry. Nikolai Ivanovich Lobachevsky (1792–1856; Figure 8.8) grew up in a poor family in Kazan, Russia, near Siberia. He, like Bolyai, was not blessed with a privileged childhood and struggled in his early years. His father, who worked as a clerk for a land surveying business, died when Nikolai was only 6 years old. However, with the support of a government scholarship, he attended the Kazan Gymnasium (Academic Secondary school) four years later. At age 15, he graduated from the Gymnasium. He was fortunate to enroll in the new Kazan State University, which was only two years old at that time, and was created as a result of the reforms of Tsar Alexander I. The students at the new University were very motivated, which made for a vibrant academic atmosphere. Nikolai was at first interested in the study of medicine. Still, at the University, he was fortunate to study mathematics under Professor Martin Bartels (1769–1833), a skilled instructor who had been encouraged to join the faculty from Germany. The history of mathematics was a favorite subject of Bartels, which he lectured on, and exposed Lobachevsky to Euclid's *Elements*, and possibly to the many disputes surrounding Euclid's fifth postulate. Bartels had taught young Gauss in school and continued to correspond with him, possibly discussing the idea of non-Euclidean geometry and the fifth postulate. Bartels appears to have nurtured Nikolai's pursuit of mathematics.

Lobachevsky rose quickly through the ranks of academia, receiving his master's degree in mathematics and physics at age 19 and rising from a lecturer to an exceptional full professor at the age of 30. He taught not only mathematics but also physics and astronomy, presenting very lucid lectures, allowing weaker students to better comprehend the concepts. Lobachevsky then began to effectively serve in the administration of the school and was appointed Rector of Kazan University which was established five years later, in 1827.

Figure 8.8 Nikolai Ivanovich Lobachevsky.

By 1826, he had formulated his ideas about non-Euclidean or hyperbolic geometry, and he submitted his paper "*A concise outline of the foundations of geometry*" to the Department of Physics and Mathematical Sciences at Kazan University. It was well received, and in 1829, he published his treatise on non-Euclidean geometry in the *Kazan Messenger*. The treatise was the first account of this departure from Euclidean geometry to appear in print. Lobachevsky replaced the fifth postulate of Euclidean geometry with this non-Euclidean geometry postulate:

There exist two lines parallel to a given line through a given point not on the line.

This postulate is illustrated on the hyperbolic surface in Figure 8.5, which shows how all the lines diverge from each other when extended. Because of the lack of communication at the time, neither Gauss nor Bolyai was aware of Lobachevsky's publication. It preceded Bolyai's Appendix by two years but was also met with skepticism among most mathematicians. Euclidean geometry was not ready to release its iron grip upon mathematics until years later.

Lobachevsky held the position of Rector for 19 years, during which time the University thrived and expanded with new buildings and programs. The level of education increased along with the quality of scientific research and new medical facilities. However, his tenure was not without its difficulties. He continued to teach a variety of topics while bearing a heavy administrative load and conducting vigorous mathematical research. This eventually took its toll on his health, and he was forced to retire from the University in 1846 at the age of 54. This was followed by the tragic death of his beloved eldest son. He was also straddled with financial difficulties, which caused his illness to become progressively worse and led to the loss of his sight. Lobachevsky's prolific mathematical achievements were given little recognition during his lifetime. His life ended in poverty in 1856 without Lobachevsky knowing the significance of his contributions and the fame he would achieve in later years. Like János Bolyai and other great mathematicians, this was another sad ending to a mathematical genius's life. Lobachevsky and Bolyai have each been recognized as having independently discovered hyperbolic geometry. Gauss was not influential in either of their achievements, and though he did not publish his ideas, he does deserve credit for his unveiling of non-Euclidean geometry. Another 30 years would have to pass for mathematicians to study and publish Lobachevsky's work and for it to be completely accepted. In the early twentieth century, it provided the mathematical foundation for Einstein's theory of Relativity.

Elliptical Geometry

Spherical geometry, discussed earlier, is a basic type of elliptical geometry and was developed by the ancient Greeks. However, it was not formally based on axioms until after the acceptance of the Bolyai-Lobachevsky geometry. The foundation for elliptical geometry emerges when the fifth postulate is changed to the following:

> *Given a straight line and a point not on the line, no lines can be drawn through the given point parallel to the given line.*

We can get a good picture of this postulate from the sphere shown in Figure 8.9.

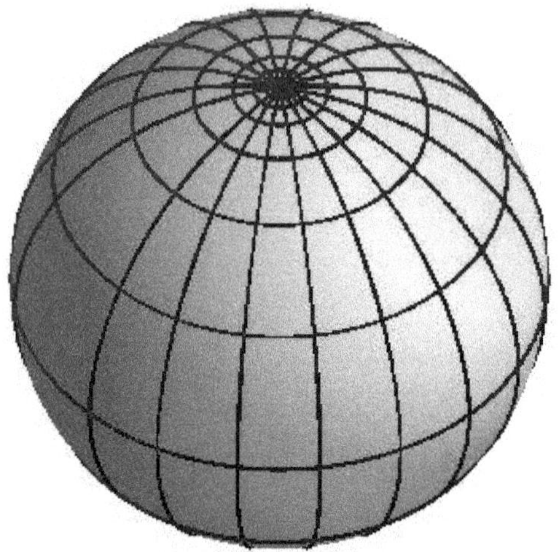

Figure 8.9   Spherical surface of constant positive curvature.

The great circles that pass through the poles are lines that divide the sphere into two halves and are the largest circles that can be drawn on the sphere. The shortest distance between any two points is the segment of the great circle drawn between the two points.

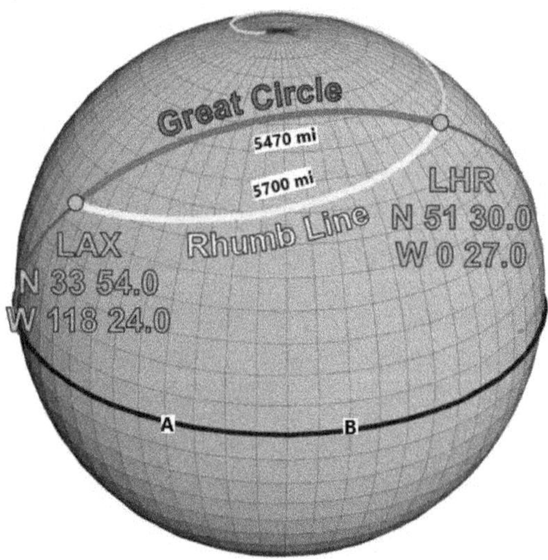

Figure 8.10   Great circle route between Los Angeles and London.

For example, in Figure 8.10, imagine traveling between the two points A and B on the equator. We can see that the shortest distance between A and B is along the equator, which is a great circle, and any other path will be longer. The great circles are therefore the "lines" or geodesics in this geometry and are analogous to the straight lines in plane geometry. Airline pilots and ship captains know that the shortest route between two locations on the Earth is the distance along the great circle drawn between the two points. Planes and ships will generally travel a great circle route to save time and fuel. Figure 8.10 also shows two routes between Los Angeles Airport (LAX) and London Airport (LHR). The *rhumb line* is the route that follows a constant compass bearing and crosses all the meridians of longitude at the same angle. The great circle route, where the compass direction continually changes, is 230 miles shorter than the Rhumb line distance as shown in Figure 8.10. For short distances, though, the compass direction or rhumb line between two points on the earth is generally followed by ships and planes because it is easier to use, and the distance is not much longer than the great circle route.

We can now see how the fifth postulate for elliptical geometry is satisfied for the special case of a sphere. Focus on the equator and all the meridians of longitude in Figure 8.10. They are all great circles, and each one constitutes a "line" in spherical geometry. Consider the equator as a given line and the north pole as a point not on the given line. Every line that passes through the pole is a meridian of longitude and *cannot* be parallel to the equator, because it intersects it. The geometry is the same everywhere on this surface, and hence this example can be shown to be true for any line and point not on the line. The fifth postulate for elliptical geometry is, therefore, true on this spherical surface.

Another example of a surface whose geometry is elliptical is the ellipsoid shown in Figure 8.11, which is formed when an ellipse is rotated in a circle. It resembles a football but is not quite pointed at the ends. The circles shown are geodesics on the ellipsoid. Most ancient civilizations very likely understood spherical geometry from their study of astronomy, which was necessary for measuring time, constructing the calendar, and predicting celestial events. The

Babylonians, the Chinese, the Mayans, the Greeks, and other early civilizations studied astronomy, but little is known about these developments.

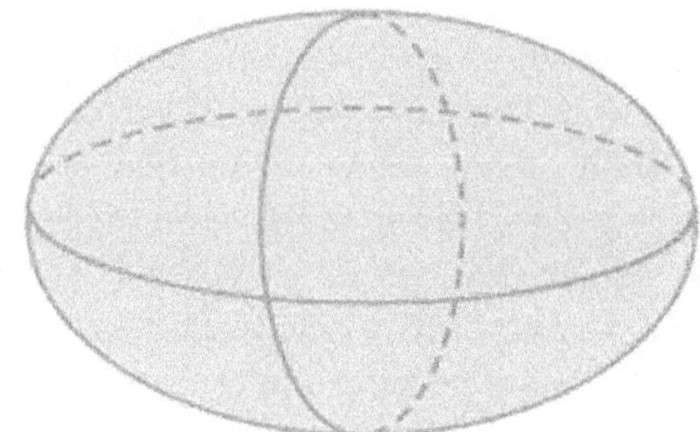

Figure 8.11 Ellipsoid.

We know that around 500 BCE Pythagoras (570–490 BCE) deduced that the earth was a sphere from his study of heavens. Euclid also included some basics of spherical geometry in Book XI of his *Elements*. The Greek mathematician Menelaus of Alexandria (70–130 CE) produced an extensive treatment of plane geometry and spherical geometry. His book *Sphaerica*, which is his only surviving book, studies arcs of great circles and how they form spherical triangles. Figure 8.12 shows a spherical triangle with three right angles where one side is the equator, and the other two sides are meridians of longitude. We define the angle where two curves meet by the angle formed by the tangents to the curves at the point of intersection. The sum of the angles of this spherical triangle is then $3 \times 90 = 270°$. In all elliptical geometric surfaces, the sum of the angles of a triangle is *greater* than 180°. In addition, the circumference of a circle is equal to $2\pi r$ for a great circle and *less* than $2\pi r$ for a smaller circle on the surface. This is opposite to hyperbolic geometry, where the sum of the angles of a triangle is *less* than 180° and the circumference of a circle is *greater* than $2\pi r$.

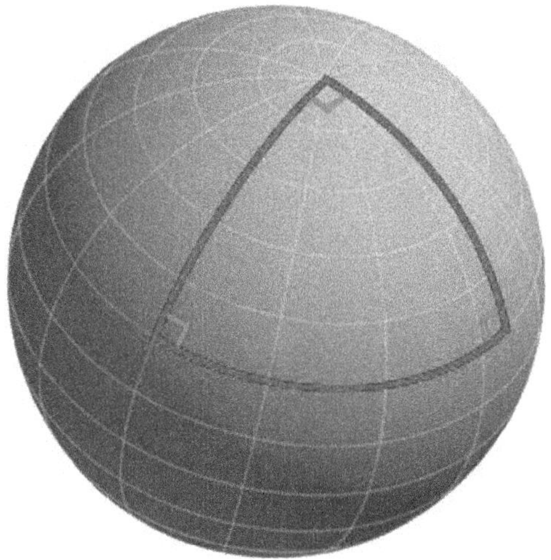

Figure 8.12   Spherical triangle with three right angles.

Many mathematicians before Bolyai and Lobachevsky, including the Persian astronomer and poet Omar Khayyam (1048–1131) and the Italian philosopher and mathematician Giovanni Girolamo Saccheri (1667–1733), in trying to prove the Euclidean parallel postulate from the first four postulates derived many theorems of both elliptical and hyperbolic geometry, but they did not envision a mathematically logical geometry different from that of Euclid. Once the curtain had been lifted by Bolyai and Lobachevsky, revealing the existence of hyperbolic geometry, prominent mathematicians of the later nineteenth century began to put non-Euclidean geometry on a firmer footing.

The Italian mathematician Eugenio Beltrami (1835–1900) was one of the first to bolster non-Euclidean geometry with his publication in 1868, *Essay on the Interpretation of Non-Euclidean Geometry*. In his publication, he produced a model of a 2-dimensional non-Euclidean geometry on the surface of a 3-dimensional figure called a *pseudo-sphere*, shown in Figure 8.13. A pseudosphere is a hyperbolic surface of constant negative curvature, like the hyperboloid in

Figure 8.5. It is formed by a curve called a tractrix that is rotated in a circle. Notice how the lines all diverge on the surface. Beltrami's model provided a final proof that Euclid's fifth postulate was essential for the foundation of his geometry, and by changing it to say that no line, or more than one line, can be drawn through a point not on a given line, one of the two non-Euclidean geometries emerges.

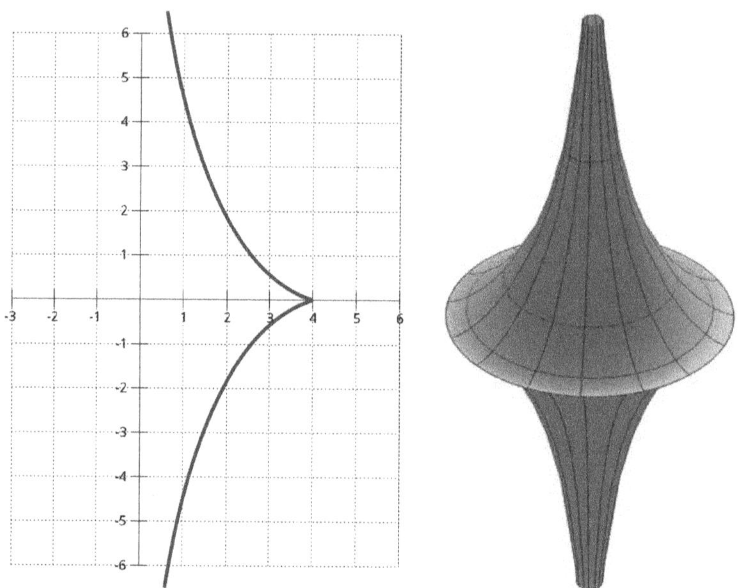

Figure 8.13   Tractrix and pseudosphere.

Geometry was lifted to a much more abstract level by the genius of the German mathematician Georg Bernhard Riemann (1826–1866). To become a lecturer at the University of Göttingen, where he was under Gauss's advisement, he gave a profound presentation in 1854 in which he proposed a very general concept of geometry. Riemann envisioned a geometric space as a structure in which one defines what is meant by length in various mathematical ways. The definition may or may not be one inspired by reality. He discussed a type of 'elliptical geometry' in which every line through a point not on a given line will eventually 'meet' the given line. Riemann's lecture was published in 1868, and though it was not fully understood until

45 years later, it further released the grip on Euclidean geometry. It paved the way for the development of a wealth of different geometries.

The year 1871 marked the definitive acceptance of non-Euclidean geometries with the work of the German mathematician Felix Christian Klein (1849–1925), who clearly showed there are essentially three major types of geometry, Euclidean, hyperbolic, and elliptic, based on the three possible cases of lines drawn through a point not on a given line. If only one parallel line can be drawn, we have Euclidean geometry. If more than one parallel line can be drawn, we have hyperbolic geometry, and if no parallel lines can be drawn, we have elliptical geometry. A surface can exhibit more than one of these geometries. Figure 8.14 shows a Torus, which is shaped like a donut and possesses all three types of curvature. The outer surface has positive curvature, while the inner surface has negative curvature, and on the very top of the Torus, where the curvature changes, there is a line of zero curvature. The General Relativity Theory of Albert Einstein (1879–1955) predicts that the universe could exhibit any one or more than one type of curvature. Space tends to have positive curvature within the attractive force of a gravitational field and could have negative curvature within a repulsive force.

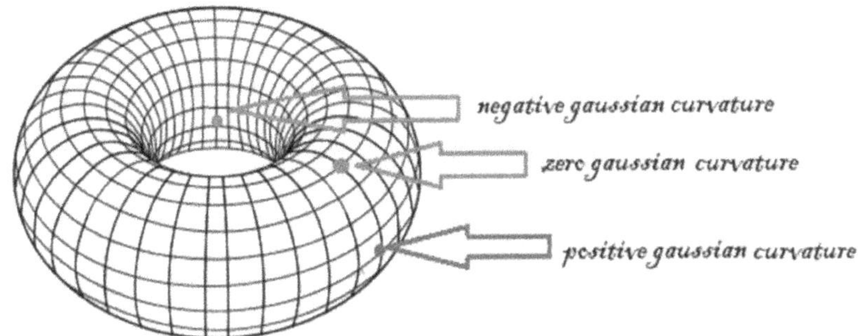

Figure 8.14   Torus.

Present evidence suggests that the universe is mostly Euclidean, with an error rate of less than 1%.

We can therefore create many different abstract geometries in the realm of mathematics based on how we define length. In our three-dimensional Euclidean world, though, there exist only three different types of geometric surfaces where length is defined in the natural way that we experience it: a Euclidean surface with zero curvature, an elliptical surface with positive curvature, and a hyperbolic surface with negative curvature. However, in the small world where we spend most of our day-to-day experience, Euclidean geometry is still the dominant force in our lives.

Chapter 9

The Theory of Sets

The theory of sets provides a basis for almost all of mathematics today. Remarkably, it is the creation of only one man, the German Russian mathematician Georg Cantor (1845–1918). This differs from many other major developments in mathematics, which tend to emerge in several brilliant minds independently and are spawned from an evolution of ideas. Georg Cantor envisioned the existence of not just one order of infinity in his theory of sets, but many orders of infinity which contradicted the belief that infinity was a unique concept. The ancient Greeks struggled with the idea of infinity, as shown in Chapter 2 with the paradoxes of the Greek philosopher, Zeno of Elea (495–430 BCE). With the discovery of calculus in the seventeenth century, the idea of an infinite sum became more conceptually understood. By the nineteenth century, infinite sets of numbers were studied in more detail and Cantor was the first to effectively attach a deeper meaning to the notion of infinity and higher orders of infinity associated with his theory of sets.

Set theory can be considered a form of mathematical logic that assigns objects into groups and subgroups, especially numbers. For example, one of the simplest sets is that of the counting numbers or natural numbers: 1, 2, 3, 4, 5, … When we add zero and the negative numbers to this set, we have the set of integers: … –5, –4, –3, –2, –1, 0, 1, 2, 3, 4, 5, … Adding fractions to the set of integers we produce the set of rational numbers. One way of looking at these sets of numbers is to consider the set of natural numbers as a subset of the integers, and the set of integers as a subset of the rational numbers.

The launch of the Russian satellite Sputnik 1 in 1957, the first to orbit Earth, sparked a Space race between the United States and the Soviet Union. In the wake of this crisis, concerns arose about the scientific status of the United States. Therefore, in 1960, the School Mathematics Study Group (SMSG) was created to examine the American school mathematics curriculum. Funded by the National Science Foundation, the initial group consisted of 10 high school mathematics teachers and 10 mathematics university professors. They were motivated to radically change the way mathematics was taught in elementary and secondary schools in the United States. The committee worked for 19 years to develop a curriculum that was referred to as the "New Math" curriculum. The theory of sets embodied such fundamental concepts that the new curriculum aimed to use set theory as a unifying idea in the teaching of mathematics to elementary students. The subject was designed to be taught as a rigorous and logical body of somewhat abstract theories rather than a series of rote concepts and computational skills. Basic arithmetic was introduced by building numbers from sets rather than rote counting. It included such topics as inequalities, number bases other than 10, symbolic logic, and abstract algebra. The SMSG developed curriculum materials and wrote textbooks, which were used to teach the New Math. After some time, the program proved too difficult for elementary students to fully comprehend and did not reinforce computational skills. Criticism from parents and teachers resulted in it being phased out years later. However, set theory remained an integral basis of mathematics and it remains an important subject taught in high school and college today.

Georg Cantor (Figure 9.1) was born in St. Petersburg, Russia, the oldest of six children. His mother was Russian and a Roman Catholic, while his father, born in Denmark, was a German Protestant which was the religion his son was brought up with. His father was an accomplished merchant, acting as an agent for wholesale goods, and later became a successful broker in the St. Petersburg Stock Exchange. His mother was very musically gifted, while his father possessed a strong love for the arts. Having these enlightened parents, young Georg Cantor became a very accomplished violinist with an affinity for the arts.

Figure 9.1   Young Georg Cantor.

He started primary school in St. Petersburg, but in 1856, his father, whose health was suffering, moved the family to Germany seeking better weather than the bitter winters of St. Petersburg. The family first settled in Weisbaden, Germany where Georg Cantor attended the Gymnasium (lower secondary school) and then moved to Frankfurt, Germany, where Cantor enrolled in the Realschule (upper secondary school) in Darmstadt, Germany. At age 15 he graduated with distinction, displaying great skills in mathematics. For the next two years, he attended the Höhere Gewerbeschule (higher vocational school) in Darmstadt as his father wanted him to become a successful engineer. Cantor was never completely comfortable in Germany or studying engineering, and looked nostalgically upon his early years in Russia. However, it was his destiny to spend the rest of his life in Germany. He wanted to study mathematics, and after seeking and being granted his father's permission, he entered the Swiss Polytechnic of Zurich in 1862. Unfortunately, a year later, his father succumbed to his illness, so that Cantor became the recipient of a substantial inheritance. This enabled him to transfer to the University of Berlin and be exposed to lectures of prominent mathematicians, including Leopold Kronecker (1823–1891) and Karl Weierstrass (1815–1897), whose names adorn many analytical theorems in mathematics. Cantor was an excellent student and, in addition to his studies, he became involved with a mathematical research group, eventually serving as their president during 1864–1865. During the summer of 1866, he studied at the University of Göttingen, and in 1867, he completed his dissertation on number theory in Berlin, receiving his doctorate that year. He took a job teaching in a girls' school but soon after secured a position at the German University of Halle, where, after presenting his thesis on number theory, he received his *habilitation*[1] in 1869. Cantor was an exceptional academic and was promoted to Extraordinary Professor at Halle in 1872, and in 1879, he achieved the rank of full professor at the young age of 34.

Cantor married Vally Guttman in 1874, and though he had six children and a modest academic income, his father's inheritance helped to provide support for the family. It was in this year of his marriage that he published the revolutionary article *Ueber eine Eigenschaft des Inbegriffes aller reellen algebraischen Zahlen* ("*On a Property of the Collection of All Real Algebraic Numbers*"), which marked the beginning of set theory. As an outgrowth of his theory of sets, Georg Cantor created the concept of orders of infinite numbers or transfinite numbers.

As with many revolutionary ideas, there is always controversy surrounding them. Such was the case with Cantor's set theory. The well-known mathematician Kronecker, who had been Cantor's

---

[1] Habilitation is a degree beyond the doctorate that requires one to produce excellent research, teach, and defend a thesis. It usually leads to a professorship, which is awarded 5 to 15 years after the PhD degree, and still exists in a number of European countries today.

mentor, and other philosophers and mathematicians, were critical of his theory, in particular, his concept that there is *more than one order of infinity*. This challenged the traditional idea that infinity existed as only one entity. He did receive support from two distinguished colleagues, Richard Dedekind (1831–1916) and Karl Weierstrass. Cantor begins with the idea that the infinite set of natural numbers: 1, 2, 3, 4, 5, ... is "countable", which means that there is a clear order to the numbers and no numbers exist between successive elements. Cantor considered the set of natural numbers to possess the first order of infinity, which he called "*Countable infinity*". He used the Hebrew symbol $\aleph_0$ (Aleph null) to denote the first transfinite cardinal number, which represents a countably infinite set. He then employs the ingenious idea of bijection, which is a one-to-one correspondence between sets, to prove that other sets are countably infinite. For example, consider the set of natural numbers compared to the set of just the even natural numbers, as shown in Figure 9.2.

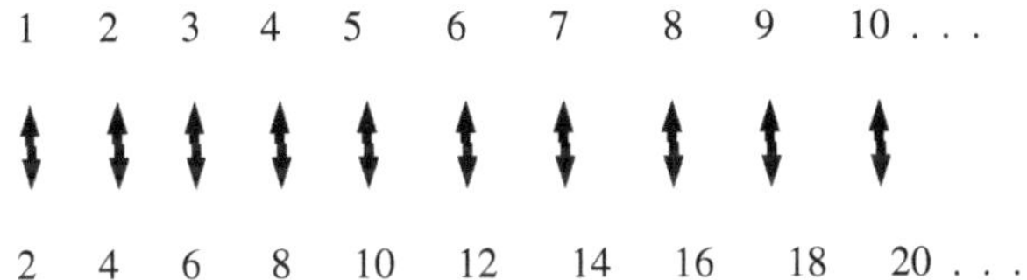

Figure 9.2   One-to-one correspondence between the natural numbers and the even natural numbers.

The arrows in the figure show that for every natural number, there is a corresponding even number. The corresponding even number is obtained by simply doubling the natural number. Therefore, there must be as many even numbers as there are natural numbers, or as Cantor would say, the set of even numbers possesses the same order of infinity, countable infinity, as the natural numbers. From this example can you show that the set of odd numbers is also countably infinite? It is possible to show many examples of countably infinite sets and some that at first seem implausible. Consider just the powers of 10, Figure 9.3 shows that we can construct a one-to-one correspondence between the natural numbers and the powers of 10.

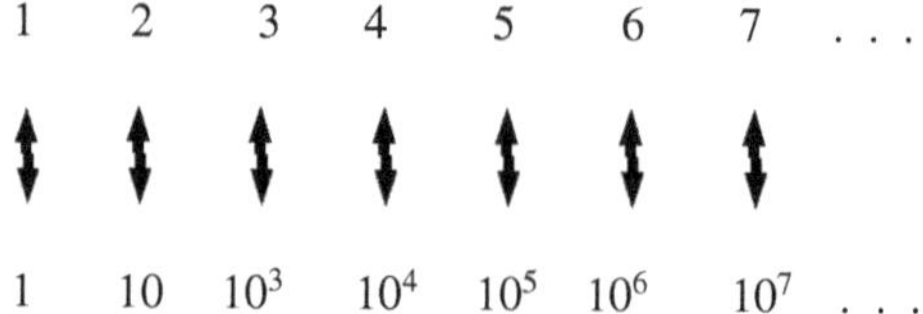

Figure 9.3   One-to-one correspondence between the natural numbers and powers of ten.

Therefore, even all the powers of ten are of the same order of infinity as the natural numbers. Each natural number is equal to the exponent of the corresponding power of 10. As inconceivable as this example and the previous one might appear, think of them as just a set of ordered symbols, without meaning, that go on forever. Viewing it this way, the logic should seem more plausible.

The infinite set of integers, which contains all the positive and negative whole numbers and zero, is also countable as shown by the one-to-one correspondence in Figure 9.4. We simply start with zero and alternate positive and negative integers:

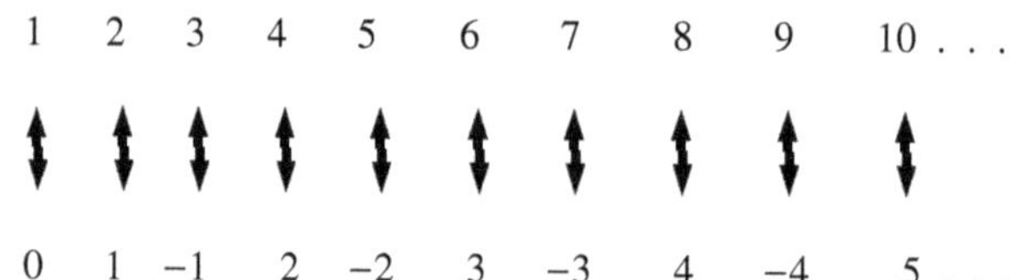

Figure 9.4   One-to-one correspondence between the natural numbers and the set of integers.

What comes next is quite impressive. Consider the entire set of rational numbers, which includes all the integers and all the fractions. At first, this set might seem to be of a higher order of infinity than the natural numbers because one can find an infinite number of fractions between any two fractions. For example, consider just two rational numbers 1 and $\frac{1}{10}$. We can keep adding fractions between these two numbers by simply adding zeroes to the denominator of $\frac{1}{10}$, that is, $\frac{1}{100}, \frac{1}{1000}, \frac{1}{10000}$, and so on. So it would appear that all rational numbers could not be arranged in any countable order. However, in 1873, Cantor showed a way of counting the positive rational numbers by placing them in the table, as shown in Figure 9.5, where the numbers of the rows are the numerators and the numbers of the columns are the denominators. This table then contains all the positive fractions, and if we now count them along diagonals as shown by the arrows in the table, one can arrange all the fractions in a one-to-one correspondence with the natural numbers. The fractions that are crossed out in the table are not counted because they reduce to fractions already counted. Therefore, it shows that all the positive rational numbers are of the same order of countable infinity as the natural numbers. Furthermore, suppose we include zero and the negative fractions in the same way as we did with the integers in Figure 9.4. In that case, we can start with zero and place the negative of each fraction after its positive value. This allows us to mathematically state that the "infinite total of fractions or rational numbers is no greater than the infinite total of natural numbers." Certainly, something to ponder about and realize why these ideas might have been controversial.

	1	2	3	4	5	6	7	8	...
1	$\frac{1}{1}$	$\frac{1}{2}$	$\frac{1}{3}$	$\frac{1}{4}$	$\frac{1}{5}$	$\frac{1}{6}$	$\frac{1}{7}$	$\frac{1}{8}$	...
2	$\frac{2}{1}$	$\frac{2}{2}$	$\frac{2}{3}$	$\frac{2}{4}$	$\frac{2}{5}$	$\frac{2}{6}$	$\frac{2}{7}$	$\frac{2}{8}$	...
3	$\frac{3}{1}$	$\frac{3}{2}$	$\frac{3}{3}$	$\frac{3}{4}$	$\frac{3}{5}$	$\frac{3}{6}$	$\frac{3}{7}$	$\frac{3}{8}$	...
4	$\frac{4}{1}$	$\frac{4}{2}$	$\frac{4}{3}$	$\frac{4}{4}$	$\frac{4}{5}$	$\frac{4}{6}$	$\frac{4}{7}$	$\frac{4}{8}$	...
5	$\frac{5}{1}$	$\frac{5}{2}$	$\frac{5}{3}$	$\frac{5}{4}$	$\frac{5}{5}$	$\frac{5}{6}$	$\frac{5}{7}$	$\frac{5}{8}$	...
6	$\frac{6}{1}$	$\frac{6}{2}$	$\frac{6}{3}$	$\frac{6}{4}$	$\frac{6}{5}$	$\frac{6}{6}$	$\frac{6}{7}$	$\frac{6}{8}$	...
7	$\frac{7}{1}$	$\frac{7}{2}$	$\frac{7}{3}$	$\frac{7}{4}$	$\frac{7}{5}$	$\frac{7}{6}$	$\frac{7}{7}$	$\frac{7}{8}$	...
8	$\frac{8}{1}$	$\frac{8}{2}$	$\frac{8}{3}$	$\frac{8}{4}$	$\frac{8}{5}$	$\frac{8}{6}$	$\frac{8}{7}$	$\frac{8}{8}$	...
⋮	⋮								

Figure 9.5   One-to-one correspondence between the natural numbers and the set of rational numbers.

We still have other important numbers to consider, and those are the irrational numbers, which are numbers that cannot be expressed as fractions. An irrational number is an infinite decimal whose digits have no pattern which includes roots of numbers that are not rational such as the square root of 3 or the cube root of 2, trigonometric values such as the sin 10_0 and many other types of numbers. Rational numbers (fractions) are infinite decimals that have a repeating pattern of digits such as $\frac{1}{2} = \mathbf{0.500000}\ldots, \frac{1}{9} = \mathbf{0.111111111}\ldots, \frac{1}{7} = \mathbf{0.42857142}$ $\mathbf{85714\ 2857}\ldots$, and $\frac{1}{11} = \mathbf{0.0909090909\ldots}$

Irrational numbers such as the square root of 2 and π, when expressed as decimals, show no pattern in their digits: $\sqrt{2} = \mathbf{1.414213562\ldots}$ and $\pi = \mathbf{3.141592654\ldots}$ In our incredible digital age, π has been calculated to 202,112,290,000,000 digits, which is over 100 trillion digits, a new world record that took 100 days to calculate. Absolutely no pattern of digits has ever been found.

Cantor tried to prove that the set of all the rational and irrational numbers, called the set of real numbers, is countable. This proved too difficult, and he went on to prove in a paper in 1874 that they are *not* countable. No matter how close two real numbers are, there is always an uncountable infinite set of numbers between them, and Cantor's method of proof implies a higher order of infinity. The set of real numbers therefore possesses a greater infinite order than the set of natural numbers. Cantor used the symbol $\aleph_1$ to denote a transfinite number that is the measure or cardinality of the real numbers. His work is not only important mathematically, but philosophically as well. As a result, strong objections to his ideas not only came from prominent mathematicians but also from well-known philosophers. Cantor was described as a scientific charlatan and a renegade. His work seemed to some Christian theologians to challenge the nature of God, which was believed to possess a unique and absolute infinity. This hostile attitude of Cantor's contemporaries took its toll on Cantor's health and thwarted his ambitions. He desired a chair at the prestigious University of Berlin, but Kronecker, who headed the mathematics department at the University of Berlin, was one of his fiercest critics. He was instrumental in preventing Cantor's application for a post at the University and made it difficult for him to leave Halle.

Figure 9.6   Older Georg Cantor.

The Swedish mathematician, Magnus Gösta Mittag-Leffler (1846–1927) was one of the first mathematicians to support Cantor's theory of sets. Cantor wrote over 50 letters to Mittag-Leffler attacking Kronecker and suffered a mental crisis, which caused him to lose confidence in his own work. Fortunately, he overcame the crisis a year later and continued his work with transfinite sets. However, he never surpassed the genius of his earlier discoveries. He believed that mathematics should be open to any creative ideas as long as there are no contradictions, and they are based on previously accepted concepts. In 1897, Cantor published a major treatise on set theory that differs little from modern texts on the subject. The year 1897 proved extremely rewarding for Cantor as the first International Congress of Mathematicians was held in Zurich, and Cantor's work was considered a great contribution to mathematics. He received much praise and was held in high esteem by prominent colleagues. He went on to establish the existence of transfinite numbers whose cardinality is even greater than $\aleph_1$. He showed that the set of all the subsets of a given set, called the *power* set, has a higher infinite order than the original set. This implies that the set containing all the subsets of the real numbers has a cardinality greater than $\aleph_1$. Though the next transfinite number after $\aleph_1$ would be $\aleph_2$, it is not established if the cardinality of the power set of the real numbers is $\aleph_2$. Cantor's theory of sets still leaves unanswered questions and some paradoxes. One is called the continuum hypothesis, which states that there are no cardinal numbers between $\aleph_0$ and $\aleph_1$. Cantor struggled to prove this hypothesis, but without success, and it continues to be an open question today. In 1899, Cantor raised another paradoxical question: "*What is the cardinal number of the set of all sets?*" He realized that it must be the greatest possible cardinal number, but no satisfactory answer has been found. There are many unanswered questions and unsolved problems in mathematics, and even some contradictions, but they do not serve to limit the power of mathematics. On the contrary, they provide constant challenges to clarify concepts and find solutions. In the case of set theory, there is little concern with its concept, and it remains a well-established foundation for much of mathematics today.

Georg Cantor's health, like that of many geniuses, suffered in his later years. From 1884, he experienced bouts of depression and wrote:

> *I don't know when I shall return to the continuation of my scientific work. At the moment I can do absolutely nothing with it and limit myself to the most necessary duty of my lectures; how much happier I would be to be scientifically active, if only I had the necessary mental freshness.*

When Cantor suffered these periods of depression, he turned his interest to philosophy and his belief that Francis Bacon wrote Shakespeare's plays as a result of an intense study of Elizabethan literature. The loss of his mother in October 1896 and his younger brother in January 1899 created great strain, and he applied for a leave from teaching during the winter semester of 1899–1900. Further tragedy struck with the death of his youngest son in December 1899, causing his struggles with depression to stay with him until the end of his life. He continued to teach but needed to take several leaves from his teaching and spend some time in a sanatorium. It is believed his mental state was due to bipolar disorder, but was not helped by the hostility he endured from colleagues. In 1911, he was very pleased to be invited to the University of St. Andrews in Scotland to attend the 500th anniversary of the founding of the University. In 1912, he was awarded the honorary degree of Doctor of Laws by St. Andrews University, but he was too ill to receive it in person, and his behavior began to become eccentric at times. He stopped teaching in 1913 and had to survive with little food during the difficult years of the first World War. His illness progressed, and in 1917, he entered a sanatorium for the last time and died of a heart attack the following year. Even after Cantor's death, there was still criticism of his work by Ludwig Wittgenstein (1889–1951), a philosopher who worked on the foundations of mathematical logic. He wrote that mathematics is "ridden through and through with the pernicious idioms of set theory." David Hilbert (1862–1943), an outstanding and well-respected mathematician of the 19th and 20th centuries, overcame much of this criticism when he praised Cantor's work as:

> *the finest product of mathematical genius and one of the supreme achievements of purely intellectual human activity.*

Chapter 10

The Digital Age

As the twentieth century moved into its second half, the rise in technology accelerated and continued into the twenty-first century at an ever more bewildering pace. It leaves many of us trying to keep up with the continuous transformations. The twentieth century spawned the vacuum tube in 1904 which gave way to radio transmission and the first electronic computer called the ENIAC (Electronic Numerical Integrator and Computer) which was built by the US army in 1945. This was followed by the first transistor in 1947, integrated circuits in 1958, portable electronic calculators and desktop computers in the 1960s, cellphones in the 1970s, and the internet and smartphones in the 1990s. Our present century has given rise to many new technologies such as Bluetooth, 3D printing, autonomous vehicles, synthetic organs, artificial intelligence, and, on the horizon, quantum computing. Many of these developments could not have occurred without the advancement of mathematical programming for computers.

The pioneers of mathematical programming emerged in the nineteenth century. In 1804, one of the earliest programmable devices was the Jacquard loom, shown in Figure 10.1, which was patented by the Frenchman Joseph Marie Jacquard (1752–1834). It was programmed using punch cards, which were attached to the loom and provided a continuous sequence of operations. Each card provided the program for one row of the woven design. As the cards were fed into a box, a combination of rods passed through the holes, causing hooks to grab a pattern of threads that were woven into the loom.

Figure 10.1   Jacquard loom.

The use of punch cards to control a process in a machine is what inspired the English mathematician Charles Babbage (1791–1871) to design the prototype of a computer. He was not only a proficient mathematician but also a mechanical engineer, an inventor, and a philosopher who can be considered the "father of the computer." Babbage designed in intricate detail two Difference Engines, which were high-powered mechanical calculating machines, and an Analytical Engine, which had all the qualities of today's electronic computers. Though he did not manage to completely construct any of his calculating machines, 153 years after he designed his Difference Engine #2, the London Science Museum spent 17 years constructing it and completed it in 2002. See Figure 10.2. It is truly a mechanical marvel, weighing five tons, able to compute numbers containing 31 digits, raise mathematical expressions to as high as seven powers, and print the results.

Figure 10.2   Charles Babbage and his Difference Engine #2.

By the time Charles Babbage was 19 and entered Trinity College in Cambridge, he had taught himself a significant amount of mathematics. He became a fellow of the Royal Society in 1816 but had little success securing a professorship at various colleges. He helped found the Royal Astronomical Society in 1820 and was awarded its Gold Medal for the invention of a small engine for calculating mathematical and astronomical tables. As a result, he received a grant to build a large difference engine with a printer, but by 1827, the money became scarce, and tragedy struck with the loss of his father, his wife, and two children, all within one year. He received a substantial inheritance from his father, which left him independently wealthy, and he continued working on his Difference Engine while receiving more grant money. Babbage was finally issued a professorship in 1828 at Cambridge, after being denied the position three times. By 1834, the grant money had run out, and he was unable to complete his large Difference Engine. At this time, he had completed the first drawings of his Analytical Engine, the precursor of today's computer.

Meanwhile, during the previous year, 1833, Augusta Ada King (1815–1852; Figure 10.3), the only legitimate child of the famous romantic English poet Lord George Byron and Anne Isabella Milbanke, attended one of Babbage's lectures and became very interested in his work, particularly the design of his Analytical Engine. Lord Byron had left his wife one month after Ada was born, and Lady Byron was concerned about her daughter's upbringing. She inculcated Ada's interest in mathematics and logic, not wanting her to develop her father's perceived insanity. Ada experienced bouts of illness in her childhood, but still studiously pursued her studies. She was only 18 years old in 1833 and a very gifted mathematician and writer when she began a long collaboration with Babbage. In 1835, she married William King, who became the Earl of Lovelace, thereby making Ada the Countess of Lovelace. She thought of herself as pursuing "poetical science" and described herself as an "Analyst and Metaphysician." Her scientific interests brought her in contact with noted scientists of the day, including the English physicist Michael Faraday and the author Charles Dickens.

Babbage's design for the Analytical Engine contained logical components very similar to today's computers. It contained a *mill*, which is the analogue of a computer's central processing unit, which operates on all the variables. Its input consisted of punch cards, which is the same type of input used by early electronic computers of the 1960s. The *store* was a memory that held 1,000 numbers of 50 digits each but was capable of infinite storage by outputting data and storing it on punch cards. Ada recognized that the Analytical Engine could follow instructions and wrote the first algorithm designed for a machine. She realized that code could be developed and repeated, which is known as a loop in programming. Along with an article about Charles Babbage's Analytical Engine, she published notes describing a step-by-step procedure for solving a mathematical problem analogous to a flowchart in today's terms. Mathematics was Ada Lovelace's gift, earning her the title "mother" of coding. She is often considered the world's first computer programmer. Her vision of computers went beyond mere calculating, which Babbage and many others envisioned as the principal capabilities of the machines. She wrote about the impact of technology on society as a creative tool. A programming language, Ada, named after her, was first designed in 1980 by a team led by Dr. Jean Ichbiah at Honeywell-Bull in France for the United States Department of Defense. It has since gone through several iterations, and today it is a high-powered language and an international technical standard.

Figure 10.3   Ada King, Countess of Lovelace.

Figure 10.4   Herman Hollerith.

As we approach the twentieth century, we find the gifted son of German immigrants, Herman Hollerith (1860–1929; Figure 10.4), working as a statistician for the United States Census Bureau, investigating ways to analyze and manipulate large amounts of data efficiently. He studied the punch card system of the Jacquard loom and, in 1884, while working for the United States Patent Office, he applied for his first patent to convert information on punched cards into electrical impulses which would activate mechanical counters. The counters used the binary system, which only employs two digits, zeros and ones to count. In our decimal system the place values from right to left are 1, 10, 100, 1000, etc., which are powers of ten: 10^0, 10^1, 10^2, 10^3, etc. In the binary system, the place values from right to left are 1, 2, 4, 8, 16, etc., which are powers of 2: 2^0, 2^1, 2^2, 2^3, 2^4, etc. Figure 10.5 shows how the first ten decimal numbers are represented in the binary system. Every number can be written in the binary system, though it uses many more digits than the decimal system. For example, 99 in the decimal system is written in the binary system as 1100011. Hollerith's cards were punched using a typewriter-like device and fed through a card reader where spring-loaded pins passed through the holes. If a pin passed through a hole, it closed an electrical circuit and activated a "1" on a dial-like counter; otherwise, a zero was activated on the counter. (See Figure 10.6.)

Hollerith gradually improved his electromechanical tabulating machine, and it was first tested in 1887 on mortality statistics in Baltimore and then tested in New Jersey and New York. Eventually, he perfected his system to handle data very quickly and won a competition for it to be used in the 1890 United States Census. The data was processed in three months by the end of 1890, compared to two years if done manually. The total population in the United States that year was tabulated at 62,622,250. This was a great achievement by

Decimal	1	2	3	4	5	6	7	8	9	10
Binary	1	10	11	100	101	110	111	1000	1001	1010

Figure 10.5   First ten binary numbers.

Figure 10.6   Hollerith's tabulating machine.

Hollerith, who gained international recognition. It marked the beginning of semiautomatic data processing using binary code that dominated data processing for nearly a century. The system was used again for the 1891 census in Canada, Norway, and Austria, and later for the 1911 census in the United Kingdom. In 1911, Hollerith's Tabulating Machine Company joined other companies to form the Computing-Tabulating-Recording Company. Thirteen years later, the company changed its name to "International Business Machines" (IBM).

The advent of World War II triggered the development of computer science, artificial intelligence, atomic energy, and quantum mechanics. Alan Mathison Turing (1912–1954; Figure 10.7), known as the "Father of modern computing," was a British mathematician, computer scientist, logician, cryptanalyst, and biologist. He was a 1934 graduate of the University of Cambridge and earned his doctoral degree in mathematical logic from Princeton University in 1938. At Cambridge, he produced his design of the Turing Machine, which remarkably describes a modern computer even though the technology did not exist at the time to construct it. His major contribution was to the war effort at Britain's Government Code and Cypher School at Bletchley Park in 1939 to 1945. He designed and built an electromechanical computer, the

Figure 10.7   Alan Turing and the Turing Machine.

Bombe, and with another mathematician, Gordon Welchman (1906–1985), helped decipher the code settings of the German Enigma machine, aiding many Allied victories. His design advanced the concepts of computation and algorithms that have provided much of the basis for modern theoretical computer science.

By 1946, the first electronic programmable computer, the ENIAC (Electronic Numerical Integrator and Computer), shown in Figure 10.8, was built and operated by the United States Army. It contained 18,000 vacuum tubes, the precursor of today's transistor. A vacuum tube contained a cathode, employing an element, like thorium, which when heated emitted negatively charged electrons. A positive plate, the anode, collected them while their flow was controlled by one or more positively charged grids. Therefore, a vacuum tube allowed varied control of the electric current. The ENIAC also contained 7,200 crystal diodes, 1,500 relays, 70,000 resistors, 10,000 capacitors, approximately 5,000,000 hand-soldered joints and weighed 30 tons. It was a complete Turing Machine whose speed at the time was one thousand times faster than existing electromechanical machines. It cost $487,000 (equal to $8,539,000 in 2025) to construct and was in continuous operation until 1955. Today, we have a microprocessor smaller than a grain of rice that has the same computing power as the ENIAC!

Figure 10.8   (a) Vacuum tube. (b) The ENIAC: Electronic Numerical Integrator and Computer.

Figure 10.8 shows an army officer setting the switches on one of ENIAC's function tables. It was believed that the ENIAC did more calculations in ten years than all of humanity had done until that time. Many very talented women made considerable contributions to computer programming and, for many years, received little credit for their important work. Kathleen McNulty Antonelli (1921–2006; Figure 10.9) was an Irish-born American and one of the few women who excelled in mathematics at Chestnut Hill College for Women in Philadelphia, graduating in 1942. At the time, the army was looking for mathematics majors to work on the ENIAC. It was being constructed at the Moore School of Electrical Engineering at the University of Pennsylvania by the physicist John Mauchly (1907–1980) and the electrical engineer John Eckert (1919–1995). Antonelli was accepted and trained at the Moore School, which employed over 70 women to do the massive calculations. She was one of six chosen women who became the first computer programmers of the ENIAC. However, when the ENIAC was finally presented at the University of Pennsylvania with a test run that greatly impressed the media, the work of the two women responsible for its success and the other four women who were instrumental in its programming was never acknowledged. These women were the very first computer programmers anywhere, but published photographs show only men working on the massive machine. Kathleen McNulty Antonelli went on to marry the physicist John Mauchly, and recognition finally came to the "ENIAC Six" a half century later when they were inducted into the Women in Technology International Hall of Fame in 1997. Credit is also due to the British computer scientist and mathematician Kathleen Booth (1922–2022; Figure 10.9), who first created assembly language and produced the "Booth multiplier algorithm" with her husband, Andrew Donald Booth (1918–2009).

The construction of modern computers, not unlike Babbage's design, has had enormous influence on the whole of mathematics. It is not an exaggeration to say that they have modernized the entire world.

The esteemed mathematician John Von Neumann (1903–1957; Figure 10.10) was born in Budapest, Hungary, and displayed remarkable talent as a young child. He mastered classical Greek at the age of 6 and could memorize a page in the phone book by simply studying it two or three times. World War I did not interrupt his education, and at age 18, he completed the Lutheran Gymnasium (high school). It was difficult for a Jewish student in Budapest at that time, so Von Neumann studied chemistry at the University of Berlin, and at the same time excelled in mathematics examinations at the University of Budapest, but he did not attend classes there.

Figure 10.9   Kathleen McNulty Antonelli (left) and Kathleen Booth (right).

Figure 10.10   John Von Neumann.

In 1926, he received a diploma in chemical engineering from the Technische Hochschule in Zürich. He continued to pursue mathematics, receiving a doctorate in mathematics from the University of Budapest the same year. His work and publications contributed to his recognition as a genius, and he achieved worldwide fame in mathematics during his twenties. He became a lecturer at Princeton on quantum theory and was appointed a professor there in 1931. In 1933, he was one of the six original professors forming the new Institute for Advanced Study, which included Albert Einstein. Among his many significant contributions to computer science was his Von Neumann architecture, which enables programs to be stored in computer memory. His work in artificial intelligence provided a basis for further research and the effective use of computers today. There were many other brilliant minds that contributed to the state of modern computer programming. Here are some of the outstanding mathematicians and computer scientists who have made significant contributions to the field of mathematics and computers.

Figure 10.11   John Warner Backus.

John Warner Backus (1924–2007; Figure 10.11) earned a master's degree in mathematics in New York at Columbia University, and while working at IBM in the 1950s created Fortran (short for **for**mula

translation). This computer language revolutionized software at that time. The language simplified programming by translating commands into the more complex machine language used by the computer. Within two years, over half of all programs were written in Fortran, earning John Backus the title "Father of Fortran." One of his reasons for his creation lies in his statement:

My work has come from being lazy, I didn't like writing programs, so I started work on a system to make it easier to write them [sic].

Figure 10.12   Ida Rhodes.

Among the outstanding women mathematicians who made significant contributions to computer programming was the Ukrainian mathematician Ida Rhodes (1900–1986; Figure 10.12). Rhodes designed the C-10 language in the early 1950s for the Census Bureau's UNIVAC I computer and created the original computer program used by the Social Security Administration. She is also known for her important development of computer programs to translate languages, in particular Russian, her native language, to English. In 1949, the

Department of Commerce awarded her an Exceptional Service Gold Medal for "significant pioneering leadership and outstanding contributions to the scientific progress of the nation in the functional design and application of electronic digital computing equipment."

Figure 10.13   Grace Brewster Hopper.

The United States Navy rear-admiral Grace Brewster Hopper (1906–1992; Figure 10.13) earned a Ph.D. in both mathematics and mathematical physics from Yale University in 1930 and became a professor of mathematics at Vassar College. During World War II, she left Vassar and joined the Navy Reserve, where she worked with a team on computers, becoming quite proficient in the art of programming. After the war, she joined the Eckert–Mauchly Computer Corporation and developed the computer language COBOL (acronym for **co**mmon **b**usiness-**o**riented **l**anguage), which is still being used today in various businesses. The United States Navy destroyer USS Hopper was named after her, and she was awarded 40 honorary degrees from universities around the world. She was a member of the Navy Reserve from 1943 until 1986 and became a rear admiral. Hopper was posthumously awarded the Presidential Medal of Freedom in 2016 by President Barack Obama.

Figure 10.14   Dorothy Jean Johnson Vaughan.

Dorothy Jean Johnson Vaughan (1910–2008; Figure 10.14) was an American mathematician and human computer who worked for the National Advisory Committee for Aeronautics (NACA) and the National Aeronautics and Space Administration (NASA) at Langley Research Center in Hampton, Virginia. She became supervisor of the west area computers and the first African-American woman to receive a promotion and supervise a group of staff at the center. This group consisted of African-American female mathematicians who performed crucial calculations for aerospace engineers. During her 28-year career, Vaughan prepared for the introduction of computers in the early 1960s by teaching herself and her staff the Fortran programming language. Her calculations and analysis were essential to the success of the early U.S. space program, including the launch of astronaut John Glenn into orbit. The 2016 film "Hidden Figures" portrays her as Dorothy Vaughn, a brilliant mathematician who broke barriers in STEM (Science-Technology-Engineering-Mathematics).

Seymour Roger Cray (1925–1996; Figure 10.15), a mathematician and an electrical engineer born in 1925 in Chippewa Falls, Wisconsin, deserves mention here as one of the pioneers of the digital age. He was called the "father of supercomputing," having designed many computers that were the fastest in the world for many years. His company, Cray Research, created the supercomputer industry and is responsible for the design of many high-performance computers used today. Seymour Cray has also been called the "Thomas Edison of supercomputers."

In 1966, Joseph Weizenbaum (1923–2008; Figure 10.16), a German American mathematician, computer scientist, and MIT professor who escaped Nazi Germany in 1936, created the first chatbot.

Figure 10.15   Seymour Roger Cray.

Figure 10.16   Joseph Weizenbaum.

He envisioned it in the role of a psychotherapist, where a user would send a message via a mainframe computer, and a party could then respond accordingly. He called it "Eliza", after Eliza Doolittle in George Bernard Shaw's play *Pygmalion*. It was the first program that allowed conversations between humans via machines. Though it provided the seed for today's artificial intelligence, Weizenbaum was adamant that it must never let machines replace humans.

Figure 10.17   Sir Tim Berners-Lee.

The person who has had one of the most significant impacts on our digital age was the English computer scientist Tim Berners-Lee (Figure 10.17), born in London in 1955, whose parents were both mathematicians and computer scientists. His mother and father worked on the first commercially built computer, the Ferranti Mark 1. After receiving a degree in Physics from Oxford and building a computer out of an old television set, Berners-Lee worked for a telecommunications company and helped develop typesetting software for printers. He then worked intermittently at CERN, the European

Council for Nuclear Research (in French, *Conseil Européen pour la Recherche Nucléaire*), where in 1989 he realized he could join hypertext with the Internet. On a computer, hypertext refers to text containing references to other texts that the user can readily access. "Hyper" is a mathematical term meaning an extension, as in "hyperspace." Berners-Lee was frustrated working with the web at CERN and was able to incorporate all the existing technology, hypertext, multifront text objects, and the Internet to raise the level of abstraction to a larger documentation system. As Berners-Lee puts it:

I just had to take the hypertext idea and connect it to the TCP (transfer control protocol) and DNS (Domain Name System) and — behold! — the World Wide Web.

Thus, HTTP in the prefix of an internet address stands for Hypertext Transfer Protocol, which provides the ability to send messages and data over the internet and permits users to interact with web pages, making it an essential component of the World Wide Web.

Lee went on to found the World Wide Web Consortium, served as a professor at the Massachusetts Institute of Technology, and is a member of the advisory board of the MIT Center for Collective Intelligence. He was knighted in 2004 by Queen Elizabeth II. He received the 2016 Turing Award, often referred to as the "Nobel Prize of Computing," for inventing the World Wide Web, its protocols, and algorithms.

The other person who has had one of the most significant impacts on our digital age is Geoffrey Everest Hinton (Figure 10.18), a British Canadian computer scientist and cognitive psychologist born in London in 1947. He comes from a distinguished family. His father was a prominent entomologist, his three siblings are all scholars, and he is the great-grandson of the following two accomplished mathematicians. Mary Everest Boole and her husband George Boole, who is the creator of Boolean logic, which provides much of the basis for modern computing. Mary Boole's uncle was George Everest, the surveyor for whom Mount Everest is named. Geoffrey Hinton received a degree in

Figure 10.18   Geoffrey Everest Hinton[1].

experimental psychology at the University of Cambridge and was awarded a PhD in artificial intelligence from the University of Edinburgh in 1978. After working both in England and the United States in neuroscience, in 1987, he became a fellow in the research program sponsored by the Canadian Institute for Advanced Research, studying artificial intelligence, robotics, and society. He became an expert in neural networks, and in 2004, he and his collaborators launched a new program, "Neural Computation and Adaptive Perception" (NCAP), which studies learning in machines and brains.

For his design of neural networks, Hinton and two of his colleagues received the Turing Award in 2018, and he became known as the "Godfather of AI." In 2024, two Nobel Prizes in physics were awarded to Geoffrey Hinton and John Hopfield for advances in the field of medicine consisting of machine learning technology using

---

[1] © Nobel Prize Outreach. Photo: Clément Morin.

artificial neural networks. Hinton is Professor Emeritus at the University of Toronto, having been associated with the school since 1978. Recently, he resigned from Google and has expressed concerns about the abuse of AI spreading fallacious information, the elimination of jobs due to its employment, and the lack of governmental safety guidelines, which could lead to AI systems surpassing human intelligence. We are entering an age of unpredictable outcomes, fostered by AI and another incredible development: Quantum Computing.

In 1981, Nobel Prize-winning physicist Richard Feynman (1918–1988; Figure 10.19) was one of the first to recognize the limits of classical computers and proposed harnessing quantum mechanics to operate a quantum computer. A digital computer relies on electrical pulses consisting of strings of binary digits 0 and 1. Simply put, no flow of current produces a 0, and a flow of current produces a 1. Because of this, digital processing speed has its limitations. A quantum computer, on the other hand, is almost infinitely faster than a digital computer. It derives its processing speed from the use of qubits or quantum bits. A qubit is a subatomic particle, such as an electron,

Figure 10.19   Richard Phillips Feynman.

which can occupy many states simultaneously. This means it can be a 0 or a 1 or any value in between all at once due to its perpetual motion.

This ability to simultaneously be in multiple states is called superposition, which is not easy for most of us to fathom; however, when several qubits are in superposition, an enormous number of calculations can be achieved. This phenomenon is not easily arrived at and requires circuits to be reduced to temperatures close to absolute zero, or for the electromagnetic field to operate in a vacuum chamber. Due to the extreme conditions required, quantum computers will not replace digital computers for most people. Quantum computers are presently being experimented with by large companies and promise to solve many extremely complex medical and scientific problems beyond the capability of a digital computer or the human mind.

Throughout the end of the 20th century and into the 21st century, progress was made toward quantum computing. In 2009, Yale University created the first rudimentary solid-state quantum processor using a two-qubit superconducting chip. This was the first time a solid-state device demonstrated quantum information processing. Quantum computers began to evolve, and in 2017, IBM produced a 50-qubit quantum computer that maintained its state for 90 microseconds, while in 2019, Google claimed to have achieved quantum supremacy with its Sycamore processor consisting of 53 qubits. In 2024, Google's quantum processor Willow performed a standard benchmark computation in less than 5 minutes that today's fastest computers would take 10 septillion (which is the number 1 followed by 27 zeros) years to compute. A truly mind-boggling accomplishment. As you read this, months after it was written, it's clear that there have been significant advancements in quantum computers, AI, robots, and other technologies. Given this rate of development, it is impossible to predict what our technical future will be like. We can only hope that it will serve to improve the world we live in.

Index

A

abacists, 87

abacus, 20, 24–25, 71

Abel, Niels Henrik, 150, 156–160

Abel Prize, 160

Abelian groups, 153

absolute geometry, 176–177

absolute zero, 223

abstract algebra, 150, 192

Abu Kamil, Shuja, 83

acceleration, 128, 130

acrophonic numbers (Greek), 40

Acta Eruditorum, 137

Ada (programming language), 207

Age of Enlightenment, 105

Ahmes Papyrus (Rhind Papyrus),
    26–30, 42

al-Biruni, 60

algebra
    beginnings of, 77–86
    cossic art, 99
    fundamental theorem of,
        102–103
    origin of word, 77
    symbolic, 95–96

algebraic notation, 95–96

algorithm, 207, 211, 213

al-jabr, 77–78

al-Khwārizmī, 63, 66, 77–83

al-muqābalā, 77–78

al-Qifti, 63

al-Tusi, Nasir al-Din, 84–85

al-Uqlidisi, 63

Analytical Engine, 205–207

analytic geometry, 106, 116,
    123–124

angle trisection (classical Greek
    problem), 51–52

antiderivative, 136

Antonelli, Kathleen McNulty,
    212–213

Apollonius of Perga, 47, 59, 77

approximation of π (Archimedes),
    42–43

Arab mathematicians, 62–64, 66,
    77–86

Archimedean spiral, 43–44, 99

Archimedes of Syracuse, 41–45,
    53, 58–59, 62, 77, 85, 91, 99,
    122–123, 128

area
  of circle, 51–52, 85–86
  under curves, 125, 134–136
  of parabola, 125, 134–136
Aristarchus, 45
Aristotle, 107, 127, 132
arithmetic progression, 173
Ars Magna (Cardano), 89–93
artificial intelligence (AI), 211, 214, 219–222
associativity, 150
astronomy (ancient), 1, 12, 21, 26, 34, 36, 45, 47, 63, 83, 85, 91, 100–102, 127
axioms
  Euclidean, 170–171
  group, 149–150

B
Babbage, Charles, 204–207
Babylonians, 6–18, 36, 56, 60–61
Backus, John Warner, 214–215
Barrow, Isaac, 127
Bartels, Martin, 181
base 2 (binary), 19, 21, 24, 31, 142, 208–210
base 5, 24
base 10 (decimal), 4, 6–8, 11–13, 19–21, 24, 35–36, 40, 56–58, 61–62, 73
base 12 (duodecimal), 5, 57
base 20 (vigesimal), 35–36, 57–58
base 60 (sexagesimal), 4–12, 18, 56, 61, 70
Beltrami, Eugenio, 187–188
Berners-Lee, Tim, 219–220
Bernoulli, Johann, 137, 140
bijection, 195

binary system, 19, 21, 24, 31, 142, 208–210
binomial coefficients, 84, 91
binomial theorem, 91
Bletchley Park, 210–211
Bolyai, Farkas, 174–177, 179
Bolyai, János, 174–180, 182–183, 187
Bombelli, Rafael, 93–94
Bombe, 211
Boole, George, 220
Boolean logic, 220
Booth, Kathleen, 212–213
Booth multiplier algorithm, 212
Brahmagupta, 63, 79–80
Brahmi numerals, 59
Briggs, Henry, 101–102
Buddha (Gautama), 62

C
calculating machines, 107–109, 133
calculus
  branches of, 122–123
  discovery of, 121–147
  differential, 122–124, 128–131
  fundamental theorem of, 135–136
  integral, 122–123, 134–137
  notation, 133–137
  priority dispute, 136–142
calendars, 19, 26, 35–36, 52–53, 63
  Gregorian, 53
  Julian, 52–53
  Maya, 35–36
Cantor, Georg, 191–201
Cardano, Gerolamo, 89–93, 112
Cardano–Tartaglia formula, 91
cardinal numbers, 195, 198–200

cardinality, 198–200
Cartesian geometry, 106
Census Bureau, U.S., 208–210, 215
central processing unit (CPU), 206
CERN, 219–220
Charta Volans, 140
Chevalier, Auguste, 165–167
Chevalier de Méré (Antoine
    Gombaud), 111–114
Chinese mathematics, 18–24, 57,
    61
circle
    area of, 51–52, 85–86
    circumference, 85, 122–123,
        179, 186
    great circle, 178, 184–186
    squaring the (classical
        problem), 51–52
classical Greek problems, 51–52
closed operation, 149
COBOL, 216
Codex Vigilanus, 64
coefficients, 95
Cogito ergo sum, 105
Colson, John, 131
combinatorics, 86
Commercium Epistolicum, 140
commutativity, 149–150
completing the square, 81
complex numbers, 92–94
composite numbers, 67
computer programming, 203–223
conic sections, 47, 108–109
continuum hypothesis, 200
Copernicus, Nicholas, 45, 102–103
cossists, 99
countable infinity, 195–198
counting boards (Chinese), 20, 61

counting numbers, 191
Cray, Seymour Roger, 217–218
cryptography, 154, 210–211
cube, doubling (classical problem),
    51–52
cubic equations, 16–18, 88–92
cuneiform, 3, 7, 9
curvature
    negative, 177–179, 186–190
    positive, 177–178, 184,
        189–190
    zero, 177, 189–190

D
decimal fractions, 5, 10–11, 64
Dedekind, Richard, 195
deductive geometry, 19, 39, 53–54
Degen, Ferdinand, 157
del Ferro, Scipione, 89–90
derivative, 130, 133, 136
Descartes, René, 44, 105–109,
    115–116, 123–125, 127, 133, 142
*Description of the Wonderful Canon
    of Logarithms* (Napier), 101
Devanagari numerals, 60
Difference Engine, 204–206
differential calculus, 122–124,
    128–131
digital age, 203–223
Diophantine equations, 70
Diophantus, 70, 118
Discours de la Méthode (Descartes),
    105–106
division (Babylonian method),
    13–14
division by zero, 80
Domain Name System (DNS), 220
doubling the cube, 51–52

drachma, 40
Dürer, Albrecht, 44, 95–99
Dürer's magic square, 97–99
dust board, 63

E
Eastern Arabic numerals, 64
Eckert, John, 212
Egyptians, ancient, 1, 5, 8, 23,
    25–32, 42, 57
Einstein, Albert, 75, 128, 183, 189,
    214
Elamites, 2, 32–34
Elements (Euclid), 39–40, 48–51
Eliza (chatbot), 219
ellipse, 47, 108–109
ellipsoid, 185–186
elliptical geometry, 172, 177,
    183–189
energy–mass equivalence, 75
Enigma machine, 211
ENIAC, 203, 211–212
ENIAC Six, 212
Enlightenment, Age of, 105
equations
    cubic, 16–18, 88–92
    linear, 81, 103, 142
    polynomial, 81, 102–104
    quadratic, 16, 63, 79–83, 88,
        95, 103, 154–155
    quartic, 89–90, 92, 94
    quintic, 157–158
equinoxes, precession of, 47
equivalent (≡), 153
Eratosthenes, 47
Essai sur les Sections Coniques
    (Pascal), 108–109
Euclid of Alexandria, 39–40,
    47–51, 77, 85, 91, 108, 117,
    127, 169–171, 186

Euclidean geometry, 169–172,
    176–179, 182–183, 189–190
Eudoxus of Cnidus, 42, 47–48, 50,
    84, 121–122
Euler, Leonhard, 75, 152, 154–155,
    157
Euler's equation, 75
exhaustion, method of, 42–43, 47,
    84, 122–123
exponents, law of, 99
extrema (maxima/minima), 124

F
false position, method of, 23, 29–30
Fermat, Pierre de, 112, 115–119,
    124–125
Fermat's Last Theorem, 116–119
Fermat's Little Theorem, 116
Ferrari, Lodovico, 89–91
Feynman, Richard, 222
Fibonacci (Leonardo of Pisa),
    64–73, 83, 87, 128
Fibonacci Association, 73
Fibonacci Quarterly, 73
Fibonacci sequence, 67–73
Fields Medal, 160
fifth postulate (parallel postulate),
    170–172, 176–177, 181–185
finger counting, 12, 35, 57, 62
Flos (Fibonacci), 66, 70
fluxions, method of, 130–131
force, law of, 128
Fortran, 214–215, 217
Fourier, Joseph, 164
fractions
    Babylonian (sexagesimal),
        10–11, 14, 18
    decimal, 5, 10–11, 64
    Egyptian (unit fractions),
        30–31, 57

rational numbers, 191–192, 196–198

sexagesimal, 10–11, 14

Frederick II (Holy Roman Emperor), 70

fundamental theorem of algebra, 102–103

fundamental theorem of arithmetic, 103–104

fundamental theorem of calculus, 135–136

G

Galileo Galilei, 102–103, 127

Galois, Évariste, 151, 160–168

Galois theory, 168

gambling, 91, 111–114

Gauss, Carl Friedrich, 78, 153, 172–175, 179–183, 188

geodesic, 177–179, 185

geometric progression, 44

geometry

    absolute, 176–177

    analytic, 106, 116, 123–124

    deductive, 19, 39, 53–54

    Egyptian, 28–29

    elliptical, 172, 177, 183–189

    Euclidean, 169–172, 176–179, 182–183, 189–190

    Greek, 39–51

    hyperbolic, 171–172, 177–183, 186–189

    non-Euclidean, 48, 169–190

    projective, 108

    spherical, 169, 172, 183–187

geocentric model, 45, 102

Germain, Sophie, 164

Girard, Albert, 102–104

golden ratio, 71–74, 88

golden rectangle, 73–74

Gombaud, Antoine (Chevalier de Méré), 111–114

Gough Rule (Pythagorean theorem), 18

Göttingen, University of, 174, 188, 194

gravity, laws of, 126–128, 138

great circle, 178, 184–186

Great Plague of London, 126

Greeks, ancient, 8, 17–19, 26, 39–54, 58, 121–123

Gregorian calendar, 53

group theory, 149–168

groups

    Abelian, 153

    axioms, 149–150

    of permutations, 156

    symmetry, 150

Gupta numerals, 60

H

habilitation, 194

Hardy, G.H., 118

Hebrew number system, 58

heliocentric model, 45, 102

Hidden Figures (film), 217

hieratic script, 28

hieroglyphics, 26–27, 32

Hilbert, David, 201

Hindu-Arabic numerals, 57–59, 64–67, 71

Hinton, Geoffrey Everest, 220–222

Hipparchus of Rhodes, 18, 47

Hispanus, Dominicus, 70

Hiyya, Abraham bar, 85–86

Hollerith, Herman, 208–210

Holmboe, Bernt, 157

Holy Roman Emperor, 70

Hopper, Grace Brewster, 216

HTTP (Hypertext Transfer
  Protocol), 220
Hundred Years' War, 86
Huygens, Christiaan, 133
hyperbola, 47, 108–109, 125, 178
hyperbolic geometry, 171–172,
  177–183, 186–189
hyperbolic surface, 177–179, 182,
  187–188
hypertext, 220

I
i (imaginary unit), 94
IBM, 210, 214–215, 223
identity element, 149–150
imaginary numbers, 71, 75, 92–94,
  103
incommensurable numbers, 48, 50
Indian mathematics, 45, 58–64,
  66–67, 69
indirect proof, 49–50
induction, mathematical, 110–111
inertia, law of, 127
infinite series, 49, 122
infinity, 48–50, 121–122
    countable, 195–198
    orders of, 191, 195–200
    transfinite, 194–195, 199–200
    uncountable, 198
Inquisition, Spanish, 86
integers, 149–150, 191–192,
  196–197
integral, 134–136
integral calculus, 122–123,
  134–137
integral sign (∫), 135–137
International Congress of
  Mathematicians, 199
inverse element, 149–150

irrational numbers, 42, 48, 50–52,
  69–70, 83, 198
irrigation systems, 6, 8, 52
ISBN (International Standard Book
  Number), 154
Islamic mathematics, 77–86

J
Jacobi, Carl, 159
Jacquard, Joseph Marie, 203–204
Jacquard loom, 203–204, 208
Johannes of Palermo, 70
Julian calendar, 52–53

K
Kant, Immanuel, 169–170
Kastner, Abraham, 174
Kazan University, 181–183
Keill, John, 140–141
Kepler, Johannes, 142
keraia sign, 40–41
Khayyam, Omar, 187
al-Kindl, 63
kinetic energy, 142
Klein, Felix Christian, 189
Kronecker, Leopold, 194–195,
  198–199

L
La Géométrie (Descartes), 106
Lagrange, Joseph-Louis, 155–157
Laplace, Pierre-Simon, 58–59
law of exponents, 99
laws of motion, 126–128
Layard, Sir Austin Henry, 3–4
Leibniz, Gottfried Wilhelm, 119,
  125, 132–144
Leonardo da Vinci, 87–88
Levi Ben Gerson (Gersonides), 86

L'Hôpital, Guillaume, 138
Liber Abaci (Fibonacci), 65–71
Liber Quadratorum (Fibonacci), 66
limit, 122
linear equations, 81, 103, 142
Liouville, Joseph, 167
Lobachevsky, Nikolai Ivanovich, 181–183, 187
logarithm, 99–102
logarithmic spiral, 44
loop (programming), 206
Lovelace, Ada (Augusta Ada King), 206–207
Lucas, Édouard, 69

M
magic square (Dürer), 97–99
mathematical induction, 110–111
Mauchly, John, 212
maxima and minima, 124, 137
Maya civilization, 34–37, 57–58
Menelaus of Alexandria, 186
Mersenne, Marin, 112, 139
Mesopotamia, 2, 6
method of exhaustion, 42–43, 47, 84, 122–123
method of false position, 23, 29–30
method of fluxions, 130–131
metric system, 73–74
Metric Conversion Act (1975), 73–74
Mint, Royal, 145
Mittag-Leffler, Magnus Gösta, 199
Möbius, August, 151
Möbius strip, 151–152
modular arithmetic, 152–154
modulo, 153–154
Moscow Papyrus, 26, 28–29
motion, laws of, 126–128

Mozart, Wolfgang Amadeus, 128
multiplication
    Babylonian method, 13
    Egyptian method, 31–32
    tables, 3, 13, 32, 56

N
Nagari numerals, 60
Napier, John, 99–102
Napoleon, 73
NASA (National Aeronautics and Space Administration), 217
natural numbers, 191–192, 195–198
negative numbers, 19, 21, 48, 50, 79–82, 88, 92–94
neural networks, 221–222
New Math, 192
Newton, Isaac, 119, 125–132, 136–147
Newtonian fluid, 144
Newton's laws of motion, 127–128
Nile River, 25–26
Nine Chapters on the Mathematical Art, 21
Nobel Prize, 159–160, 221–222
non-Euclidean geometry, 48, 169–190
notation
    algebraic, 95–96
    calculus, 133–137
    Leibniz's, 133–137
    Newton's, 133
number systems
    Babylonian, 6–18
    binary, 19, 21, 24, 31, 142, 208–210
    Brahmi, 59
    Chinese, 18–21, 57

Egyptian, 26–28, 57
Greek, 40–41, 58
Gupta, 60
Hebrew, 58
Hindu–Arabic, 57–59, 64–67, 71
Maya, 34–37, 57–58
Nagari, 60
Roman, 52–53, 55, 58, 64, 67, 71
number theory, 39, 48, 116, 118–119, 151–152, 172, 194

O
obol, 40
Old Kingdom (Egypt), 26
Oldenburg, Henry, 139
Omar Khayyam, 187
On the Sphere and Cylinder (Archimedes), 43
one-to-one correspondence, 195–197
Opticks (Newton), 144
optics, 126–127, 144

P
Pacioli, Luca, 87–88
papyrus
Moscow, 26, 28–29
Rhind (Ahmes), 26–30, 42
parabola, 47, 108–109, 125, 134–136
paradoxes of Zeno, 48–50
parallel postulate (fifth postulate), 170–172, 176–177, 181–185
Paris Academy, 142, 159, 163
Parthenon, 73–74
Pascal, Blaise, 84, 107–115, 119, 128, 133
Pascal's triangle, 84, 109–111

Pascaline, 108–109
perfect numbers, 116
permutations, 150, 156, 164
perspective (art), 88, 95, 99
Philosophiae Naturalis Principia Mathematica (Newton), 137–139
pi (π), 42–43, 48, 50–52, 75, 85, 95, 123, 154, 198
Picasso, Pablo, 128
Pitiscus, Bartholomeo, 102
place value
Babylonian, 8–10, 56, 60–61
Chinese, 20
Hindu–Arabic, 58–64, 66
Maya, 35–36
plague, 115–116, 126–127
Plato, 39, 47, 127, 144
Platonic solids, 50, 88, 99
Playfair, John, 171
Playfair's axiom, 171
Plimpton 322, 7, 17–18
Poisson, Siméon Denis, 166
Polo, Marco, 87
polydactyly, 57
polynomial equations, 81, 102–104
pope, 93
postulates (Euclid), 48, 170–171
power set, 199–200
Practica Geometrie (Fibonacci), 66
prime numbers, 50, 67–69, 103–104, 116
Princeton University, 210, 214
Principia (Newton), 137–139
printing press, 59, 66, 87
probability theory, 84, 91, 109–115
projective geometry, 108
proportion theory, 47–48
pseudosphere, 187–188

Ptolemy, Claudius, 45, 77, 102
punch cards, 203–204, 206–209
pyramids, 1, 26
Pythagoras of Samos, 45–46
Pythagorean theorem, 18, 33,
    45–47, 116–117
        Chinese knowledge, 18
        proof, 45–47
Pythagorean triple, 117

Q
qibla, 84
quadratic equations, 16, 18, 63,
    79–83, 88, 95, 103, 154–155
quadratic formula, 16, 63, 73,
    154–155, 158
quadrature of the circle, 51–52
quantum computing, 203,
    222–223
quartic equations, 89–90, 92, 94
qubit, 222–223
quintic equation, 157–158

R
rabbits (Fibonacci's problem),
    69–70
rate of change, 123, 128–130
rational numbers, 191–192,
    196–198
real numbers, 198–200
reciprocals, 14
reductio ad absurdum, 49
regular polyhedra, 50
relativity, theory of, 183, 189
Renaissance, 71, 82, 87–105, 123
Rhind Papyrus, 26–30, 42
Rhodes, Ida, 215–216
Riemann, Georg Bernhard,
    188–189

right triangle, 17–18, 33, 45–47,
    116–117
Roman Empire, 39, 52–53, 58
Roman numerals, 8, 52–53, 55, 58,
    64, 67, 71
Royal Mint, 145
Royal Society of London, 133,
    139–140, 145
Rudolf, Christoff, 99
Ruffini, Paolo, 156, 158
Russell, Bertrand, 49

S
Saccheri, Giovanni Girolamo, 187
Sand Reckoner (Archimedes), 43,
    45, 62
School Mathematics Study Group
    (SMSG), 192
Schumacher, Heinrich Christian,
    174
scientific method, 104
Sebokht, Severus, 62
set theory, 191–201
SI system (metric), 73–74
sine curve, 125
Sosigenes of Alexandria, 52–53
space race, 192
Spanish Inquisition, 86
sphere, surface area and volume,
    43
spherical geometry, 169, 172,
    183–187
spherical triangle, 186–187
spherical trigonometry, 84, 102
spiral
        Archimedean, 43–44, 99
        logarithmic, 44
        on shells, 44, 71
Sputnik, 192

square roots
    Babylonian method, 14–15, 18
    extraction, 14–15, 43, 69
squaring the circle, 51–52
statistics, 114
Stevin, Simon, 99
subset, 192, 199–200
Sumerians, 2–8, 11–12
Summa de Arithmetica (Pacioli),
  87–88
sunya (zero), 61
superposition, 223
symbolic algebra, 95–96
symmetry group, 150

T
tables
    Babylonian, 13–18
    multiplication, 3, 13, 32, 56
    of reciprocals, 14
    of squares and cubes, 13
    trigonometric, 47
tabulating machine, 209–210
tangent line, 124–125, 131, 137
Tartaglia (Niccolò Fontana),
  89–91
Taurinus, Franz Adolph, 180
telescope, reflecting, 144–145
Theaetetus, 47, 50
time measurement, 7–8, 26
topology, 151–152
Torus, 189
tractrix, 187–188
Traité du Triangle Arithmétique
  (Pascal), 109–110
transfinite numbers, 194–195,
  199–200

transistor, 203, 211
triangle
    area of, 29
    Pascal's, 84, 109–111
    right, 17–18, 33, 45–47,
      116–117
    spherical, 186–187
trigonometry, 17–19, 43, 47, 84–85,
  102, 104
trisecting an angle, 51–52
Turing, Alan Mathison, 210–211
Turing Award, 220–221
Turing Machine, 210–211

U
unit fractions (Egyptian),
  30–31, 57
UNIVAC I, 215
United States Metric Board, 74
University of Berlin, 194, 198
University of Budapest, 213–214
University of Cambridge, 126–127,
  130, 145, 205, 210, 221
University of Göttingen, 174, 188,
  194
University of Halle, 194, 198
University of Leipzig, 132
University of Padua, 91
University of St. Andrews, 201
University of Toronto, 222
US customary system, 73–74
USS Hopper, 216

V
vacuum tube, 203, 211
Vaughan, Dorothy Jean Johnson,
  217

velocity, 123–124, 128–130
Viete, Francois, 94–96
von Neumann, John, 212–214
von Neumann architecture, 214

W
Weierstrass, Karl, 194–195
Weizenbaum, Joseph, 218–219
Welchman, Gordon, 211
Western Arabic numerals, 64
Weyl, Hermann, 165
Wiles, Andrew, 118–119
Wittgenstein, Ludwig, 201
World Wide Web, 220

Z
Zeno of Elea, 48, 121, 191
zephir (zero), 66–67
zero
absolute zero, 223
Babylonian lack of, 10, 36, 56
Chinese use of, 20
curvature, 177, 189–190
division by, 80
Hindu–Arabic, 57, 60–61,
66–67, 69
Maya invention of, 35–37
symbol for, 10, 20, 35–37, 61,
66–67